OTHER TITLES OF INTEREST FROM ST. LUCIE PRESS

ENVIRONMENTAL MANAGEMENT TOOLS ON THE INTERNET

Accessing the World of Environmental Information

Michael Katz
Dorothy Thornton

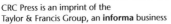
CRC Press
Taylor & Francis Group
Boca Raton London New York

CRC Press is an imprint of the
Taylor & Francis Group, an **informa** business

SOIL
AND WATER
CONSERVATION
SOCIETY

First published 1997 by St. Lucie Press

Published 2018 by CRC Press
Taylor & Francis Group
6000 Broken Sound Parkway NW, Suite 300
Boca Raton, FL 33487-2742

First issued in hardback 2018

ISBN 13: 978-1-138-46540-4 (hbk)
ISBN 13: 978-1-57444-059-1 (pbk)

This book contains information obtained from authentic and highly regarded sources. Reasonable efforts have been made to publish reliable data and information, but the author and publisher cannot assume responsibility for the validity of all materials or the consequences of their use. The authors and publishers have attempted to trace the copyright holders of all material reproduced in this publication and apologize to copyright holders if permission to publish in this form has not been obtained. If any copyright material has not been acknowledged please write and let us know so we may rectify in any future reprint.

Visit the Taylor & Francis Web site at
http://www.taylorandfrancis.com

and the CRC Press Web site at
http://www.crcpress.com

Table of Contents

How to Use This Manual

We cannot tell or show you everything there is to know about the Internet. Particularly, we cannot show you *all* the best places to access environmental information. With such explosive growth on the Internet, new and better information appears every day. This manual was out of date from the moment we wrote it.

We have therefore tried to emphasize the tools you need to find valuable information so that if, for example, a site address were to suddenly change (and they do), you would know how to go about finding it.

However, this manual does not end with the tools. It is also designed to show you a number of sites where you can pull actual data off the Internet. Often, when using the Internet, it is not a case of just finding information, but rather sifting through copious amounts of information for what is useful and relevant. This manual is intended to show you a number of sites that contain information you need in your day-to-day professional life.

To accomplish these goals, this manual is divided into four chapters.

Chapter 1 describes the equipment and programs you need to connect to the Internet. It also provides you with the basic vocabulary you will need to determine the connection that is best for you and your business.

Chapter 2 provides general information about what is on the Internet and how to access it.

Chapter 3 uses the tools and vocabulary covered in Chapter 2 to show you how to get environmental information off of the Internet. This chapter only refers to information that is available for free. Information is divided into eight categories:

- Regulatory Information
- Chemical-Specific Information
- Water Information
- Land/Soil Information
- Air Information
- Hazardous Waste Information
- Release and Risk Information
- General Environmental Information

Chapter 4 covers the database services that are available on the Internet that charge fees.

All information that an Internet user would have to enter using keystrokes has been written in **boldface.** Computer program names in text have been written in *italics* so that they can be easily distinguished. Internet addresses have also been written in *italics*.

A glossary of terms and index have been provided which include the seemingly endless number of acronyms used in the manual.

Often a computer program name has come to have several meanings. In the same way that xerox is now synonymous with photocopy and Xerox refers to the company, we use words such as telnet, archie, finger, and ftp to represent the name of a particular program, the collective names of all programs that perform the same function, and on occasion as a verb to describe the actions performed by the program. As your familiarity with the program names and their functions increases, the ambiguity of the language will not present any problems. We have tried to use the language as clearly as possible.

Appendix A contains examples of some of the data and files available through the Internet.

This manual is designed for newcomers to the Internet (newbies) and sophisticated users alike. Consequently, if you are a newbie, you are not necessarily going to understand everything that is written in Chapter 2 the first time around, particularly if you are not sitting at a computer at the time that you read it. Do not worry if this happens. As you use the Internet more, you too will become a sophisticated user.

We have made use of two icons to differentiate:

Essential newbie information

More sophisticated user information

1

What You Need to Connect to the Internet

What is known as the Internet today began as a U.S. Department of Defense Advanced Research Projects Agency (ARPA) project in the 1960's. The project's goal was to connect different computers in different locations and allow several of the computers to share communication lines. During the 1960's and 1970's, the project grew as more universities and companies, mainly defense contractors, connected to what was called ARPANet. Computers in other countries connected to ARPANet by the end of the 1970's, creating a worldwide network.

As more and more computer systems and networks were connected to ARPANet, it came to be known as the Internet in the 1980's. Discouraged by the high cost of Internet access and stringent government regulations, some companies and hobbyists learned how to connect their systems to the Internet. Some of these companies began to offer access to the public and the stage was set for the explosive growth of the Internet.

In the 1990's, inexpensive access began to be widely available. The creation of the World Wide Web (web), along with graphic user interfaces such as Macintosh and Windows, made navigation relatively simple for anybody with a computer and a modem. The Internet, which encompasses electronic mail, the web, file transfer protocol, gopher, etc., is now growing exponentially. According to Network Wizards*, from July 1995 to January 1996 the number of internet hosts increased 43% from 6,642,000 to 9,472,000.

*Information from Network Wizards at http://www.nw.com/

To connect to the Internet you need:

- *a computer*
- *a modem*
- *an Internet Service Provider (ISP)*
- *programs*

➭ 1.1 COMPUTERS

A computer runs software (programs) to talk via a telephone line connected directly or indirectly to the Internet. An indirectly connected computer is connected to another computer that is in turn connected to the Internet. A directly connected computer is itself connected to the Internet. Nearly any computer can be used to connect to the Internet. However, if you are a PC user and want graphical access to the web, you will need a computer capable of running Windows. The speed at which information is displayed on your computer is largely dependent on the speed of your modem. The most significant exception is graphic and audio information, which can be drastically affected by the speed of your computer in addition to the speed of your modem.

➭ 1.2 MODEMS

A modem connects your computer via a telephone line to the Internet through an ISP. If you have an Internet connection at work, special high-speed modems or routers are typically used.

1.2.1 Modem Speed

The speed at which information travels through the phone line to your computer depends on the capability of your modem and the modem of the ISP. The speed of modems is measured in bits per second (bps). For practical purposes you do not need to know what a bit is or that 8 of them are called a byte. You just need to know that the faster the modem speed, the faster you can transfer information from the Internet to your computer. Typical modem speeds are 14,400 bps (or 14.4k) and 28,800 bps (28.8k). Older modems have speeds of 9600 bps, 2400 bps, 1200 bps, or 300 bps. Unless you have a lot of time and no money, you should have at least a 9600 bps modem. For web access, you should use a 14,400 bps or 28,800 bps modem. A 28,800 bps modem should be compatible with the v.34 standard, which is a protocol for transmission of data at 28,800 bps. Look for this information in the modem's manual or specifications.

1.2.2 Compression

Modems have the ability to compress data that they transfer, increasing the total throughput of data. For example, a 14,400 bps modem using compression could transmit information at a rate greater than 14,400 bps by as much as 50%.

1.2.3 Error Detection and Correction

Error control protocols filter out telephone line noise and automatically retransmit data that has changed because of line noise. When your modem with an error control protocol connects to another modem with an error control protocol, the two modems are able to filter out line noise and maintain what is known as a reliable connection. The line noise is still there, but it does not appear on your screen and, more importantly, in any data you transfer. MNP 4 and V.42 are the two most common error control protocols. You may see them mentioned in your modem's manual or specifications.

1.2.4 External vs. Internal Modems

You may choose to use an external modem or an internal modem. An external modem has its own power supply and connects to the serial port or modem port of your computer. An internal modem uses your computer's power supply and plugs into a slot inside your computer. Some disassembly of your computer is required to install an internal modem, but is not beyond the abilities of most novice computer users.

The advantage of an external modem is that you can reset the modem by turning it on and off without turning off your computer. External modems also have lights that indicate the status of your modem, including whether it is connected, whether the phone is off the hook, and whether it is receiving or transmitting data.

➥ 1.3 INTERNET SERVICE PROVIDERS

Unless you or your company have your own connection to the Internet, you will be accessing the Internet through an ISP. America Online (AOL), CompuServe (CIS), and the Microsoft Network (MSN) (the Big Three) all provide access to the Internet, email, the web, newsgroups, and some but not all other services available on the Internet. In addition, these services also offer access to databases of newspapers, magazines, and other information that are not available on the Internet. These services generally bill a small monthly fee of about $10 that includes 3 to 5 hours of service plus an hourly fee for additional use. If you think you will be spending a lot of time on the Internet and do not need access to news and database services, you may want to choose a smaller ISP which can provide much cheaper access.

Other ISPs also offer direct access to the Internet. These ISPs include Netcom, Delphi, CRL, Hooked, Best, Portal, Institute for Global Communications, the WELL, and thousands of other smaller ISPs throughout the country. All Internet services are usually available through these ISPs, but they do not generally offer access to databases of newspapers, magazines, and other information. Fees for some of these services are set as discussed above. However, many of them offer unlimited access for a monthly fee of $20 to $30.

If you choose a smaller ISP over the Big Three and you want to have graphic access to the web, you will want a type of connection known as a PPP or SLIP account rather than a shell account. The difference between a shell account and a PPP or SLIP account is that with a PPP or SLIP account, your computer becomes part of the Internet, while with a shell account, you are connected to a computer that is part of the Internet. That is, a user with a shell account is connected indirectly to the Internet, while a user with a PPP or SLIP account is connected directly to the Internet. PPP and SLIP accounts are similar; PPP is just a newer protocol. This type of account will usually be somewhat more expensive than a shell account. Most smaller ISPs also provide a shell account with their PPP accounts.

You generally need to be connected directly to the Internet to be able to run web browsers such as Mosaic and Netscape (discussed in the following chapter). However, with a shell account, the ISP computer will usually have on it all the programs you might want to use, so you do not have to have copies of the programs on your own computer. If you have an older, smaller computer, you might prefer to use a shell account.

The type of connection between your ISP and the Internet will also affect the speed at which you can surf the Internet, but not nearly as much as your modem speed. T1 and T3 are leased-line connections that are able to carry much more data than normal telephone lines. For example, a T1 line can handle up to 1,544,000 bps. A T3 line can handle up to 45,000,000 bps. If speed is of paramount importance to you, check with your provider to ensure that they have enough T1 or T3 lines for the amount of traffic generated. For most large ISPs, this should not be a problem.

1.3.1 Advantages of a Big Three Account

- Ease of setup

- Convenience of use (for example, you receive one bill from the ISP provider rather than a number of bills from each pay database you access)

1.3.2 Advantages of a Non-Big Three Account

- Lower cost of use

- Ability to customize your email and web addresses

➪ 1.4 PROGRAMS

Programs tell your computer how to send and receive email, how to download programs, how to surf through the web, and how to use a computer at another location as if you were sitting in front of it.

If your service provider is one of the Big Three, then the programs you need will be provided in your start-up package. If you use a smaller ISP, you might have to provide your own programs.

For shell accounts on smaller ISPs, programs are usually on the ISP's machines, ready for your use. Programs include *pine* and *elm* for email, *tin* for reading newsgroups, *lynx* for text-based access to the web, *telnet*, *gopher*, *ftp (file transfer protocol)*, and *archie*. *Telnet*, *gopher*, *ftp,* and *archie* are discussed in Chapter 2. To access one of these programs through your shell account, just type its name.

Shareware and freeware programs for PPP and SLIP accounts are available at ftp sites throughout the Internet from which they can be downloaded to your computer. Freeware is, as you might expect, free. You are expected to pay for shareware if you like the program and use it.

Although we will tell you in this section where to find the programs on the Internet so that you can download them to your computer, we will be describing exactly how to do this only in Chapter 2, where we describe file transfer protocols.

☞ 1.4.1 Where to Find Programs

The ftp and web addresses, called URLs (for Universal Resource Locator), from which popular programs can be downloaded, are given below for both PCs and Macintoshes. These are not the only sites where the programs can be found, nor are these the only programs available. This section is intended to provide useful reference information if you need to obtain a copy of these programs. URLs should be entered without any spaces or line breaks.

PROGRAM	URL FOR A PC RUNNING WINDOWS 3.1 OR 3.11
Eudora	An email program *ftp://gatekeeper.dec.com/pub/micro/msdos/win3/* *winsock/eudor152.exe*
Trumpet Telnet	*ftp://gatekeeper.dec.com/pub/micro/msdos/win3/* *winsock/trmptel.zip*
WinFTP	*ftp://gatekeeper.dec.com/pub/micro/msdos/win3/* *winsock/winftp.zip*
Finger	*ftp://gatekeeper.dec.com/pub/micro/msdos/win3/* *winsock/cfing13.zip*
Gopher	*ftp://gatekeeper.dec.com/pub/micro/msdos/win3/* *winsock/gophbk11.zip*
Archie	*ftp://gatekeeper.dec.com/pub/micro/msdos/win3/* *winsock/wsarch08.zip*
Trumpet Winsock	A program needed to run all other PPP programs *ftp://gatekeeper.dec.com/pub/micro/msdos/win3/* *winsock/twsk21f.zip*
NCSA Mosaic	*ftp://ftp.ncsa.uiuc.edu/Mosaic/Windows/win31x* *mosaic.exe*
Netscape Navigator	*ftp://mirror.utdallas.edu/pub/web/netscape/navigator* *2.02/windows/n16e1202.exe* or *http://home.netscape.com/comprod/mirror/client_* *download.html*

PROGRAM	URL FOR A PC RUNNING WINDOWS 95
Telnet	*ftp://ftp.vandyke.com/pub/vandyke/ntcrt10.zip*
Cute FTP	*ftp://papa.indstate.edu/winsock-l* */ftp/CuteFTP.Betas/cf3214f7.zip*

PROGRAM	URL FOR A PC RUNNING WINDOWS 95
NCSA Mosaic	*ftp://ftp.ncsa.uiuc.edu/Web/Mosaic/Windows/Win95/ mosaic20.exe*
Netscape	*ftp://mirror1.utdallas.edu/pub/web/netscape/navigator/ 2.02/windows/n32e202.exe* or *http://home.netscape.com/comprod/mirror/client_ download.html*

PROGRAMS	URL FOR A MACINTOSH COMPUTER
Eudora	*http://www.umich.edu/~archive/mac/util/comm/ eudoral.51.sit.hqx*
Fetch	An ftp program *http://www.umich.edu/~archive/mac/util/comm/ fetch2.12.sit.hqx*
Archie	*http://www.umich.edu/~archive/mac/util/comm/ anarchie1.60.sit.hqx*
TurboGopher	*http://www.umich.edu/~archive.mac/util/comm/ gopher/turbogopher2.01.sit.hqx*
NCSA Telnet	*ftp://ftp.ncsa.uiuc.edu/Telnet/Mac/Telnet2.7/2.7b4/ Telnet-2.7b4-fat.sit.hqx*
NCSA Mosaic	For the Power PC Mac: *ftp://ftp.ncsa.uiuc.edu/Mosaic/Mac/ NCSAMosaic201.PPC.hqx* For older Macs: *ftp://ftp.ncsa.uiuc.edu/Mosaic/Mac/ NCSAMosaic201.68k.hqx*
NetScape	*ftp://ftp.utdallas.edu/pub/web/netscape/navigator/2.02/ mac/netscape2.02installer.hqx.z* or *http://home.netscape.com/comprod/mirror client_download.html*

2
What You Can Find
on the Internet

The Internet is so huge and its growth so explosive that any description of what can be found on it will soon be obsolete. In this chapter we will be discussing:

- *the World Wide Web*
- *electronic mail*
- *mailing lists*
- *telnet*
- *archie*
- *ftp*
- *gopher*
- *newsgroups*

The intention of this chapter is to provide you with enough tools to be able to explore the Internet on your own, so that as it changes and new information becomes available, you will know how to access that information.

⇦ 2.1 WORLD WIDE WEB

The web is not an entity unto itself, but more a way of transmitting and formatting information over the Internet. Largely because of its ability to transmit and display full-color graphics, video, and sound, this is the fastest growing portion of the Internet.

2.1.1 Web Browsers

Pages are the basic units of information of the web and may contain text, graphics, sound, or even video. While the term "page" is used, the contents of a web page may contain many pages of information. Web pages are connected to one another using "links." In order to view these pages, you must use software known as a browser, such as Netscape, Mosaic, or Lynx (a text-only browser). If you are connected to the Internet through AOL, CIS, or MSN, you likely have a browser built into the software you use to connect to those services. If not, contact customer service and they can provide you with the latest version of their software. The browser interprets and displays the pages on your computer.

In order to take advantage of the capabilities of the web, you will need to use a browser such as Mosaic, Netscape, or the browser provided by your ISP. To help speed things up, most graphic browsers offer an option to not load graphics. If you are in a rush, skip the pretty graphics — it is much faster.

2.1.2 Links

Links on a page are highlighted using underline, bold, or a different color. By selecting a link and clicking the mouse button (or pressing return in Lynx) you receive the page to which the link refers. This may be text, graphics, video, sound, a program, or some combination of all these items. If you download a program, make sure you check it for viruses.

When you click on a link, the information represented by that link is transmitted to you. For example, when you click on the *Data Resources* link on the *Hydrology Web Page* shown in Figure 2.1.2, the requesting computer (your computer or the ISP's computer if you have a shell account) attempts to establish a connection to the computer where the *Data Resources* document is stored, the host computer. After a connection is established, a request for the selected document, in this case the *Data Resources* document, is sent to the host computer. The document's filename and the path to the file are specified in the link.

The host computer then sends an acknowledgment that the request was received, and when it is ready, the host computer transmits the *Data Resources* document to you and closes the connection. Choosing another link starts the process over again.

[Keyword Search | Submit an Entry | What's New]

This list is maintained at the Pacific Northwest Laboratory's Earth and Environmental Sciences Center (EESC) by Tim Scheibe. If you would like to have your server or organization listed, please use the submission form or send e-mail to *scheibe@phoebus.pnl.gov.*

See also Clearinghouse for Subject-Oriented Internet Resource Guides

Usage statistics.

- Related Internet Resource Lists
- Data Resources
- Computer Modeling and Software
- Publications
- Bibliographic Materials
- Projects and Specific Topics
- Electronic Discussion Groups
- Scientific and Professional Societies
- Conference and Short Course Announcements
- Universities
- Other Research Organizations
- Companies
- Miscellaneous Resources

[Keyword Search | Submit an Entry | What's New]

Version 2.0 - Last Modified 01/08/96

Figure 2.1.2

In summary, every time you click on a link:

- a connection is established
- a request for a document is made
- an acknowledgment of the request is sent
- a search for the requested document is conducted
- the requested document is sent
- the connection is closed

2.1.3 Universal Resource Locators

Web pages have addresses known as Universal Resource Locators (URLs). A typical URL is *http://www.epa.gov*. The *http* stands for hypertext transfer protocol and indicates that the page is a true web page capable of graphics, sound, and video. Other possibilities include:

- *ftp://ftp.epa.gov*
- *gopher://gopher.epa.gov*
- *news://sci.environment*
- *telnet://epa.gov*.

These are not true web pages and display only text. Ftp, gopher, newsgroups, and telnet are discussed elsewhere in this manual.

After the *http://* is a domain name separated by periods, such as *www.epa.gov*. A domain is the general address where the page you are looking at is located. As with email domain names, the domain name gives you some basic information about a site, such as whether it is governmental, commercial, or educational.

After the domain name is a path description, separated by slashes, such as *~archive/mac/util/comm/*, that specifies a particular directory where files are located. Sometimes a port number, preceded by a colon, is specified before the path description. After the domain name, a filename such as *eudora.51.sit.hqx* may be specified.

In summary, a URL usually consists of:

- a site-type identifier (http, ftp, telnet, gopher, etc.)
- a domain name
- a path
- a filename

2.1.4 Search Engines

There is no official directory of web pages, but there are quite a few places that have attempted to catalog the content of the web. You look through the web using a search engine which is, essentially, a catalog of the information available on the web. Some of the search engines on the web are:

Alta Vista	[http://www.altavista.digital.com]
Yahoo	[*http://www.yahoo.com*]
Lycos	[*http://www/lycos.com*]
WebCrawler	[*http://webcrawler.com/*]
EINet Galaxy	[*http://www.einet.net*]
Internet Sleuth	[*http://www.intbc.com/sleuth*]
All-in-One Search	[*http://www.albany.net/allinone/*]
SavvySearch	[*http://guaraldi.cs.colostate.edu:2000/form*]

To use a search engine, type in the URL address listed above in the appropriate box in your browser. Then follow the instructions for the search engine. This usually means typing in the subject or keyword that you are interested in or clicking on an area of interest.

Figure 2.1.4 is a printout of the *Yahoo* search engine home page.

The *All-in-One Search* page provides access to all of these search engines, and more. *SavvySearch* provides the ability to query many of these search engines simultaneously. These search engines have databases of web pages that you can use to search for a subject that interests you. These databases may not be comprehensive or up-to-date. Use as many of them as you can to ensure that you find the information on your area of interest, if it exists on the web.

Yahoo! Quick Access + Web + [Win a trip to Super Bowl(tm) XXX!] Launch

Options

☐ **Arts**
Humanities, Photography,
Architecture, ...

☐ **Business and Economy [Xtra!]**
Directory, Investments,
Classifieds, ...

☐ **Computers and Internet**
Internet, WWW, Software,
Multimedia, ...

☐ **Education**
Universities, K-12, Courses, ...

☐ **Entertainment [Xtra!]**
TV, Movies, Music, Magazines,
Books, ...

☐ **Government**
Politics [**Xtra!**], Agencies,
Law, Military, ...

☐ **Health**
Medicine, Drugs, Diseases,
Fitness, ...

☐ **News [Xtra!]**
World [**Xtra!**], Daily, Current
Events, ...

☐ **Recreation**
Sports [**Xtra!**], Games, Travel,
Autos, ...

☐ **Reference**
Libraries, Dictionaries, Phone
Numbers, ...

☐ **Regional**
Countries, Regions, U.S. States, ...

☐ **Science**
CS, Biology, Astronomy,
Engineering, ...

☐ **Social Science**
Anthropology, Sociology,
Economics, ...

☐ **Society and Culture**
People, Environment, Religion, ...

Text-Only Yahoo + Contributors

Figure 2.1.4.

2.1.5 Home Pages

A home page is analogous to the title page of a book or the entrance hall of a building that lists the names of the building tenants. Home pages are generally named quite logically, so you can guess a company's or organization's home page address. For example, if you want to find Blymyer Engineers' home page, try *http://www.blymyer.com*. Or if you are looking for the U.S. Department of Transportation's web page, try *http://www.dot.gov*.

As with email addresses (discussed later), web pages based in other countries will generally have two letters at the end of the URL that indicate the country. See page 17 for a discussion of email addresses.

Some home pages are visually impressive. For example, Figure 2.1.5 shows the second home page screen of the National Institute for Water Resources.

2.1.6 How It Works

Click on your web browser (Netscape, Mosaic, AOL, etc.) icon, which should (if it has been set up right) connect you to your Internet account. When the connection is established, the web browser will load up what is called your home page. This is a default web page usually determined by your browser or your ISP. For example, Mosaic usually defaults to the Mosaic home page at the University of Illinois.

From there, if you are interested in finding out information about a particular subject, all you have to do is type in the address of one of the search engines, and then ask it to search for the subject of interest to you.

➯ 2.2 ELECTRONIC MAIL

Email is, by far, the Internet resource that most people have access to. Every ISP provides at least this basic form of Internet access; that is, they provide sufficient space to send and receive email.

Figure 2.1.5

If you have a PPP or SLIP account, you will need to obtain email software, such as Eudora, yourself.

Some ISPs, such as CIS, make it easy to send and receive email from other people using the same ISP, but may make it quite difficult to send email to or receive email from someone who has an Internet account with a different Internet provider.

2.2.1 Email Addresses

An email address consists of a *user name* and a *domain name*. The *user name* is often the first letter of the person's first name followed by that person's last name. The *domain name* is the name of the company or organization providing the Internet connection. The email address of a person is written as *user_name@domain_name*. The domain name may be a series of names separated by points or numbers separated by points. For example, Blymyer Engineers' email address is *blymyer@blymyer.com*. Notice that there are no spaces in the email address. Spaces are generally not allowed in email addresses. An underscore character _ or slash / usually substitutes for a space.

2.2.2 What You Can Tell from an Email Address

The domain name of the email address contains information that you can use to determine some information about where the email came from.

For example, most U.S. email addresses end in three letters, such as

> *.com* (commercial)
> *.gov* (government)
> *.org* (organization)
> *.edu* (educational institution)
> *.mil* (military)
> *.net* (network)

Email addresses from other countries end in two letters which are an abbreviation for the country. Some examples of these are:

ar - Argentina	au - Australia	at - Austria
be - Belgium	br - Brazil	ca - Canada
cl - Chile	cr - Costa Rica	cu - Cuba
cz - Czech Rep.	dk - Denmark	do - Dominican Republic
ec - Ecuador	fi - Finland	fr - France
de - Germany	gr - Greece	hk - Hong Kong
ie - Ireland	il - Israel	it - Italy
jp - Japan	lt - Lithuania	lu - Luxembourg
my - Malaysia	mx - Mexico	na - Namibia
nl - Netherlands	nz - New Zealand	no - Norway
pe - Peru	ph - Philippines	pl - Poland
pt - Portugal	pr - Puerto Rico	ru - Russian Federation
sg - Singapore	si - Slovenia	za - South Africa
es - Spain	se - Sweden	ch - Switzerland
tw - Taiwan	th - Thailand	uk - United Kingdom
us - United States	uy - Uruguay	su - USSR (former)
ve - Venezuela		

If you have the email address *jdoe@berkeley.edu*, you would know that the email came from an educational institution known as berkeley, also known as UC Berkeley.

Computers, as a rule, are very exacting. If you make the slightest mistake in an email address, the email will either be delivered to the wrong person or, more likely, it will be returned as undeliverable.

2.2.3 Email Address Directories

While directories of web pages are pretty well established, directories of email addresses are less comprehensive and there are not very many of them.

A company, Four11, claims to be the Internet's largest email white page directory, with 6.5 million listings. The service is free and is at:
> *http://www.four11.com/*

Aldea Communications also provides an Internet white pages at:
http://www.aldea.com

The Massachusetts Institute of Technology has a gopher with links to directories at numerous companies and universities at:
gopher://sipb.mit.edu:70

Many professional organizations are also beginning to include email addresses in their membership directories.

If you know that a person is at a certain place, you may be able to figure out a likely email address. For example, if John Doe works for NASA, you might try to send email to *jdoe@nasa.gov*. If you want to check the email address prior to sending a message, you may be able to use a program called *finger* which is discussed below.

☞ 2.2.4 Finger

Finger is a program used to get information about a user or find out who is logged into a system. It is most frequently used as follows:

> finger *user_name@domain_name*
> For example, **finger mkatz@blymyer.com**

This might return the following information:

> blymyer.com
> Login name: mkatz   In real life: Mike Katz
> Directory: /home/user/mkatz   Shell: /bin/shell1
> Affiliation: None   Home System: blymyer.com
> Last login Thu Oct 15 11:43 (PDT) on ttyp6 from blymyer.com
> No unread mail
> Project: None

To find out who is logged into a system, you could use finger as follows:

> finger *@domain_name*

For example, typing **finger @blymyer.com** will tell you who is logged into the Blymyer system.

You cannot finger accounts held by some ISPs such as CIS and AOL.

☞ 2.2.5 Sending Files Through Email

In the 1960's, a standard for the 128 codes used to represent numbers and letters was developed so that different computers could exchange information. That standard, the American Standard Code for Information Interchange (ASCII), is now used whenever two computers wish to exchange information.

When you type a message in an email program, the email program stores the message as ASCII text, not as a binary file. Thus the email message can be sent to whatever email address you specify.

Most word processing programs allow you to save a file as an ASCII document. When you do this, your formatting commands are not converted into ASCII, only the text is so a plain text unformatted file (no bold, italic, tabs, or indents) will be received.

If you tried to simply send a word processing file through email without first saving it as an ASCII file, it would probably be completely unintelligible to the person receiving the file, if it even got that far.

Any file, including text files, program files, data files, etc., can be sent via email; it is just a little complicated to do it. To send an intact file (a file along with all its formatting commands) via email in most systems, you first must convert everything in the file which is represented as binary data, to ASCII text. There are programs available to convert files to ASCII text that can be transmitted in an email message. Two examples of such programs are *BinHex* (Macintosh) and *Uuencode* (PC). After one of these programs has converted a binary file into ASCII text, the ASCII text can be loaded into an email message and sent. The recipient of the email must use a decoder such as *BinHex* or *Uudecode* to convert the ASCII text back into the original binary file.

If your email system is MIME-aware (MIME=Multipurpose Internet Mail Extensions), you can attach files to messages and send them to other users without worrying about encoding and decoding. Of course, the person you send the message and file to must have a MIME-aware email system as well.

2.2.6 Security Concerns with Receiving Files

While email messages are currently safe from viruses, encoded files or files attached to messages may contain viruses. Check encoded files for viruses before and after decoding **AND** before you run the file or access it with another program.

2.2.7 Email Privacy

You should consider any email sent over the Internet equivalent to a postcard. Email sent over the Internet travels via many computers; some public, some private. At any one of these machines, your email might be intercepted and read by someone.

☞ If you want to send a letter in the equivalent of an envelope, you must use some form of encryption to protect the contents of the message. A software program known as Pretty Good Privacy (PGP) can be used freely for personal use. If you want to use it for your business, you must buy a version of the program from a company called ViaCrypt in Phoenix, Arizona at:

http://www.viacrypt.com

Note that this program may not be exported out of the U.S. or Canada, although there are some limited provisions for its use by subsidiaries of U.S. companies outside of the U.S.

2.2.8 Mailing Lists

Mailing lists offer two big advantages over every other method of getting information: speed and convenience. Messages and information are delivered to your email inbox. The information comes to you instead of you needing to go look for information. Most mailing lists are moderated (someone can control which messages come through) so there is not as much noise (irrelevant or inflammatory messages).

Mailing lists may feature lively discussions of current issues or may only become active when a "hot" issue is discussed.

To join (or subscribe to) a mailing list, you send an email message to the mailing list administrative address which is usually *listserver@domain_name* or *majordomo@domain_name*. The subject of the message is left blank and the message body includes a command such as:

> subscribe <listname> <your name or email address>
> Ex: **subscribe EPA-TRI mkatz@blymyer.com**

If your message was successful, you will receive some acknowledgment in the form of an email message welcoming you to the mailing list or a message stating that your request succeeded. If that happens, email from the mailing list called "EPA-TRI" will be sent to the address "mkatz@blymyer.com".

Mailing lists can be very difficult to find. Good places to start are newsgroups that are related to the subject you are looking for and Publicly Accessible Mailing Lists at *http://www.neosoft.com/internet/paml*. Mailing lists at this site are indexed by name and subject. The full list of Publicly Accessible Mailing Lists can be found at *ftp://rtfm.mit.edu/pub/usenet/news.answers/mail/mailing-lists*.

☞ The most common problem that newbies to mailing lists have is mistaking the administrative email address (where you send the message to join the list) and the mailing list address (where you send messages to everyone who is on the mailing list).

The two main types of mailing lists are *ListServer* and *Majordomo*. Although they both do pretty much the same things, they do them in different ways and offer different capabilities.

Requests can be made of either mailing list program. The requests should be written in the body of an email addressed to the administrative address of the mailing list, namely, *listserver@domain_name* or *majordomo@domain_name*.

If you do not understand all the request options described below, do not worry about it. The commands you need to understand are **subscribe**, **unsubscribe** and **help**. The rest will start to make more sense as you become more familiar with using mailing lists.

Here is how it works:

- everything appearing in [ ] below is optional
- everything appearing in < > is mandatory

2.2.8.1 ListServer

Here is a brief description of the set of requests recognized by ListServer mailing lists. Not every ListServer mailing list will offer these options.

All arguments (the contents within the brackets) are case insensitive (you can use lower or upper case as you wish). The vertical bar ("|") is used as a logical OR operator between the arguments. Requests to the mailing list may be abbreviated, but you must specify at least the first three characters.

Recognized requests are:

subscribe <list> <your name>
        Ex: **subscribe EPA-TRI Mike Katz**

This subscribes you to the specified list.

unsubscribe <list>
signoff <list>
        Ex: **unsubscribe EPA-TRI**

Either of these messages can remove you from the specified list. Do not try to unsubscribe someone else from a list. It is not appreciated by the administrator of the list and can result in you being taken off the mailing list.

help [topic]
Ex: **help search engine**

This will give you specific information on the selected topic. Topics may also refer to requests. If you do not list a topic and only type **help**, then the mailing list will return information about this topic. To learn more about this system, issue a **help listserver** request. To get a listing of all available topics, generate an error message by sending a bogus request like **help me**.

☞ **set <list> [<option> <arg[s]>]**

If you only type **set**, you will get a list of all current settings for the specified list. Issue a **help set** request for more information.

☞ **recipients <list>**
☞ **review <list>**

Both these commands give you a listing of all non-concealed people subscribed to the specified list.

☞ **information <list>**

Gives information about the specified list. The precise information provided depends on the list. Most often the information includes who the administrator of the list is, the posting address for the list, some guidelines for the list, and what the list is about.

☞ **lists**

Gives you a list of all local mailing lists that are served by the same administrative address, as well as all known remote lists.

☞ **index [archive | path-to-archive] [/password] [-all]**

Gives you a list of files in the selected archive (place where old messages are stored), or the master archive if no archive was specified. If an archive is private, you have to provide its password as well.

☞ **get <archive | path-to-archive> <file> [/password] [parts]**

Get the requested file from the specified archive. Files are usually split in parts locally, and in such a case you will receive the file in multiple email messages. An "index" request tells you how many parts the file has been split into and their sizes; if you need to obtain certain parts, specify them as optional arguments. If an archive is private, you have to provide its password as well.

☞ **search <archive | path-to-archive>] [/password] [-all] <pattern>**

Search all files of the specified archive (and all of its subarchives if -all is specified) for lines that match the pattern. The pattern can be a regular expression with support for the following additional operators: "~" (negation), "|" and "&" (logical OR and AND), "<" ">" (group regular expressions). The pattern may be enclosed in single or double quotes.

☞ **release**

Get information about the current release of this ListServer system.

☞ **which**

Get a listing of mailing lists to which you have subscribed that are served by the same administrative address.

2.2.8.2 Majordomo

Most Majordomo mailing lists understand the following commands:

subscribe <list> [<address>]
> Ex: **subscribe EPA-TRI** or **subscribe EPA-TRI mkatz@blymyer.com**

The first example will subscribe whatever account you are writing from to the list. If you specify an address, that address will be added to the specified mailing list.

unsubscribe <list> [<address>]
> Ex: **unsubscribe EPA-TRI dthornton@blymyer.com**

☞ **get <list> <filename>**

Allows you to get a file related to the specified mailing list.

☞ **index <list>**

Returns an index of files you can retrieve for the specified list.

☞ **which [<address>]**

Finds out which lists you, or the address you specified, are on.

☞ **info <list>**

Retrieves the general introductory information for the named list.

☞ **lists**

Shows the mailing lists served by the Majordomo server.

☞ **help**

Retrieves a message that usually includes the available commands.

☞ **end**

Stop processing commands (useful if your email program adds a signature).

☞ **who**

Returns a list of people, usually just their email addresses, who are subscribed to the list.

➪ 2.2.9 Netiquette

Sooner or later you will want to ask a question or post a message to a newsgroup or mailing list. Before you do that, spend some time (days or weeks) reading messages and getting accustomed to acceptable protocol for the newsgroup or mailing list.

The Internet, like all communities, has its own customs that you are expected to observe. If you forget these customs, you will likely be made aware of your transgression by what is known as a flame. A flame is generally an unkind personal attack that chastises you and points out the error of your ways.

Here are some tips to avoid the most common mistakes:

Upper case messages

Use upper and lower case when sending or posting messages. ALL CAPITAL LETTERS is interpreted as shouting and is hard to read.

Replying to messages

When replying to a message, do not include the entire text of the message in your reply. Only include that portion of the message that you are addressing in your reply so that your reply is given some context.

On the other hand, do include enough of the message to which you are replying so that people know what you are talking about. Messages like "Me too" are particularly useless and annoying.

Privacy

Do not post all or part of a message that was sent to you privately to a mailing list or newsgroup.

When replying to mailing lists, check the "TO:" and "CC:" lines in the header of the message to make sure the message is going where it is intended.

The official word of your company

Avoid saying things that could be misconstrued as representing your business officially.

Particularly in the consulting business: disclaim, disclaim, disclaim...

Email humor

If you are trying to be funny or use satire, do not assume that anybody else will think so. It is hard to interpret the tone of typewritten messages.

Use emoticons or smileys like :-)
[Turn your head sideways to see the face]

Do not assume that you are speaking to a U.S. audience — you are not

Terms like "un-American," "unconstitutional," and "apple pie" do not have the same impact on a worldwide audience that they might have in America.

Do not send lines longer than 70 characters

Lines with more than 70 characters may cause wrapping on some systems, making your message difficult to read. Using 70 characters or less also allows space for reply indicators.

Read the FAQ

In response to the perpetual questions of new users (newbies), many mailing lists and newsgroups offer a list of Frequently Asked Questions, known as a FAQ.

Advertising

Do not post blatant advertisements unless you know that it is acceptable or you are willing to accept the consequences, which can range from mail bombs (so much email sent to your account that it causes your account or your ISP to shut down) to annoying phone calls and faxes. Posting a message to dozens or hundreds of unrelated newsgroups may be prohibited by your ISP.

2.2.10 Common Abbreviations

The widespread use of the Internet has made typists out of many people who either prefer not to type or cannot type very well. In the interest of brevity, useful abbreviations have come into common usage. The following is a sampling of some of the most common abbreviations:

BTW   by the way
IMHO   in my humble opinion
LOL   laughing out loud
ROTFL   rolling on the floor laughing
RTFM   read the [expletive deleted] manual
TIA   thanks in advance
YMMV   your mileage may vary

☞ 2.3 TELNET

Telnet is a way to use another computer system as if you were directly connected to that computer. The word itself is both a program name and a general name for a category of programs that allow you to remotely login to another computer. Telnet can be used to access libraries or databases maintained by the government, universities, or other organizations that allow public access. In some cases, you may be required to enter a login name and password to access a computer using telnet. During the time you are logged into the other system, the commands you are familiar with may not work.

A telnet site is usually identified by a domain name such as *glis.cr.usgs.gov* or four numbers separated by periods such as *128.206.1.3*. Telnet addresses may be converted to URLs for use by your web browser by adding *telnet://* to the address so that it looks like *telnet://glis.cr.usgs.gov*.

2.3.1 How Telnet Works

To telnet to a site from a web browser, type in the URL and then follow instructions. For example, type **telnet://198.137.188.2**

To telnet from a shell account, type **telnet telnet.address**. For example, to telnet to the Seattle Public Library Federal Register database, type in **telnet.198.137.188.2**. Then follow instructions.

To telnet from a PPP account, click on the telnet icon. Generally, the telnet program will prompt you to enter a telnet address. Then follow instructions.

☞ 2.4 ARCHIE

Archie is a database that catalogs programs stored at more than 1,000 sites around the world. The system periodically checks the file libraries and catalogs the contents. You can reach archie by telnet or email.

You can access archie by telnetting to any one of the following sites:

archie.sura.net
archie.unl.edu
archie.ans.net
archie.rutgers.edu

If a login name is requested, type **archie**. Then just follow the instructions that appear on your screen.

You can also use archie by email. Send an email message to:
archie@quiche.cs.mcgill.ca

Leave the subject blank. Inside the body of the message, type **prog** followed by the filename or partial filename. For example: **prog netscape** as the body of your email asks archie for the location of the program *Netscape* available on the Internet. You can then go to one of the sites identified and download the program.

You can also run archie from your computer if you have a PPP or SLIP connection, but you need to have the archie software. You can download the software as listed on pages 7 and 8 of Chapter 1.

When using archie from a shell account, the most important archie command to remember is "prog". To look for a file, type **prog** followed by the filename (or a part of the filename, if you are unsure of its complete name).

How to download a computer program is described in File Transfer Protocol, below.

☞ 2.5 FILE TRANSFER PROTOCOL

Ftp is one of the easiest ways to transfer files via the Internet. The term refers to both a program name and the general class of programs that transfer files.

2.5.1 Types of Files

There are many types of files stored in computers. Files may contain text, graphics, databases, programs, video clips, etc. Most filenames have standard extensions that indicate what type of file they are and how they are stored. These extensions are generally found after a "." at the end of a filename. For example, *winftp.exe*, *fetch2.12.sit.hqx*, and *wsarch08.zip*.

2.5.1.1 Programs

.Zip is one of the most common filename extensions. Files with *.zip* have been compressed using a compression program called *pkzip* or *winZip* and are intended for use on PCs. In order to decompress (or uncompress) these files, you must use a program called *pkunzip* or *winZip* on the program you have downloaded. When offline, type **pkunzip** *filename*; for example, type **pkunzip winftp.zip**. Make sure you check the program for viruses before you run it. Another common method of compressing files is denoted by the *.lzh* filename extension. Use the *Lha archiver* to decompress *.lzh* files.

Files that end with *.exe* are programs that can be run on PCs as soon as they are downloaded. *.exe* stands for executable files.

Hqx is another of the most common filename extensions. Files that end with *.hqx* have been converted to ASCII with *BinHex*, a program used for Macintosh computers. *BinHex* or *StuffIt* can be used to decode the file.

.Sit filename extensions denote files compressed with *StuffIt*, which can also be used to decompress the file. *.Sit* files are mostly for Macintosh computers.

Files that end with *.sea* are programs that can be run on Macintosh computers as soon as they are downloaded. *.sea* stands for self-extracting archive files.

Filename extensions for Unix files include *.Z* or *.gz* for compressed files and *.tar* for tape archive files.

2.5.1.2 Text

Text files sometimes, but not always, have filenames that end with a *.txt* or *.asc* extension. A filename with *.html* indicates that it contains text for a web page.

2.5.1.3 Graphics

Of the many graphics and picture formats available, the most common are denoted by *.gif* or *.jpg*. Other popular formats are denoted by *.pcx*, *.tif*, *.eps*, *.cdr*, and *.cgm*. The Adobe Acrobat format, denoted by *.pdf*, is popular for organizations that want to ensure the look and feel of a document. The Internal Revenue Service (IRS) makes its forms available in *.pdf* files and the Government Printing Office (GPO) makes the Federal Register (FR) available in *.pdf* files.

2.5.1.4 Sound

Filenames for files containing sound are generally denoted by *.au*, *.wav* or *.mov*

2.5.1.5 Video or Animation

Video or animation clips are usually denoted by *.avi*, *.mpg*, *.mpeg*, *.fli.*, or *.mov*

2.5.2 How ftp Works

If you are using a web browser, most ftp file transfers occur transparently. Just tell your browser to go to the ftp site address and which file you want. For example, if you want to download NCSA Telnet for your Macintosh computer, you can do this by typing in the URL of the file you wish to download and then hitting enter. For example, type:

ftp://ftp.ncsa.uiuc.edu/Telnet/Mac/Telnet2.7/2.7b4/Telnet-2.7b4-fat.sit.hqx

You will be prompted to tell the web browser where to put the file it downloads, or the web browser will send the file to a default directory.

You can download a file using an ftp program without a web browser; however, this manual does not explain the procedure.

If you do not have access to ftp, you can still get files, but it requires significantly greater effort.

First, send an email message with a blank subject line to one of the ftp mail servers at any of the addresses listed below:

ftpmail@decwrl.dec.com
ftpmail@sunsite.unc.edu
bitftp@pucc.princeton.edu

The message of the text is used for commands that will tell the ftp mail server where to get the file and what to do with it. Here is an example of a message that would retrieve a binary file (all programs are binary files):

reply thornton@crl.com   (tells the ftp site who to send the file to)
connect ftp.ncsa.uiuc.edu   (tells the ftp site the domain you wish to connect to)
binary   (tells the ftp site what kind of file you are looking for)
uuencode   (tells the ftp site to uuencode the message)
get Telnet/Mac/Telnet2.7/2.7b4/Telnet-2.7b4.fat.sit.hqx
                                                    (gives the path to the file you are looking for)
quit   (tells the ftp site that is all you want to do)

Note that there would be no blank lines in the actual message sent through the email. It would look like this:

reply thornton@crl.com
connect ftp.ncsa.uiuc.edu
binary
uuencode
get Telnet/Mac/Telnet2.7/2.7b4/Telnet-2.7b4.fat.sit.hqx
quit

The uuencoded file will then arrive in your email-box. Save or export the message to your local disc drive or hard drive. You will need to uudecode the file and check it for viruses before you can run it. These steps are usually conducted offline (when no longer attached to the Internet).

Open your uudecoding program and uudecode the file you just downloaded. Now you can use a virus-scanning program to ensure that the downloaded file does not contain any viruses.

➭ 2.6 GOPHER

Gopher is a program used to "tunnel" for information stored in places all over the Internet. In that it is only text-based, it is a more primitive form of the web. However, a lot of information is available only via gopher. For example, the National Toxicology Program (NTP) database containing abstracts of toxicological studies was not available on the web as of September 1995; however, the gopher site can be accessed with a web browser. It was and is available through gopher. Gopher is so named not only because it "tunnels," but also because it was written at the University of Minnesota, whose mascot is a gopher.

Gopher usually begins with the menu of the University of Minnesota gopher. The menu is a selection of files and directories that can be selected by highlighting the desired selection and clicking the mouse or pressing return. If you highlight a directory, then another menu appears. If you highlight a file, then the contents of that file are transferred to your screen. Like the web, the selections on the menu may "link" you to another file or directory on the same computer or at a computer in another location.

Gopher sites have addresses that usually look like: *gopher.epa.gov*. Gopher site addresses may be converted to URLs so that they can be read by your web browser by adding *gopher://* in front of the address so that they look like: *gopher://gopher.epa.gov*.

2.6.1 Searching Through Gopherspace

Gopherspace is the term used to describe the collection of information that is stored on gopher servers throughout the world. *Veronica* is a way of searching for key words in gopher titles on more than 99% of the world's gopher servers and is accessed via gopher. A *Veronica* search will return a gopher menu of titles that match the query.

Most gopher programs default to the University of Minnesota gopher at *gopher.tc.umn.edu.* From the main menu, select "Other Gopher and Information Servers" and then select "Veronica."

You can also connect to the veronica home menu at:
gopher://veronica.scs.unr.edu:70/11/veronica

Veronica understands the logical operators AND, NOT, OR, and parentheses, which are used to order searches. If you use multiple words in your search, such as "hazardous waste," it is the same as "hazardous AND waste." Be careful with the OR operator; its use will generally result in thousands of hits.

An asterisk "*" can be used as a wildcard to match anything and must be used at the end of words.

Words used for a veronica search must be at least two characters long or they will be ignored.

Veronica interprets the query order from the right side of the query. If you are in doubt about how a search will be interpreted, use parentheses.

2.6.2 How Gopher Works

Generally all you have to do is click on your gopher icon and then follow instructions that appear on your screen.

Note that although you can turn a gopher site into a URL by adding *gopher://*, you cannot gopher to a URL that begins with anything other than *gopher://*. That is, you can use your web browser to reach gopher sites, however you cannot gopher to a web (http), ftp, or news site.

From a shell account, simply type **gopher** at the prompt and then follow instructions.

2.7 NEWSGROUPS (Usenet)

Newsgroups are areas of the Internet where you may post information and messages so that anyone can read and respond to them. Newsgroups are a part of Usenet, which is just a collection of machines that exchange information posted to newsgroups. The best analogy for newsgroups is to think of Usenet as a collection of cork boards where people put up messages. People reply to messages by attaching their message to the message to which they are responding. This collection of messages regarding the same subject is referred to as a thread.

There are more than 13,000 newsgroups covering such diverse topics as David Letterman's Top Ten List, Adobe Pagemaker, and hydrogeology. Each newsgroup has a name that consists of words separated by periods and, sometimes, dashes. Example of newsgroup names include: *ba.environment*, *sci.geology*, and *soc.culture.mexico*.

Newsgroups are organized in a hierarchy with eight major categories:

alt   Alternative topics that do not really fit anywhere else. These groups tend to have very lively and somewhat unorganized interactions. Examples include newsgroups for fans of celebrities.

comp   Computer and related topics of interest including software, hardware, computer science, and computer protocols.

misc   Miscellaneous topics. The misc category is very similar to the alt category and includes newsgroups on law and activism.

news   Topics concerning the news network (Usenet) and newsgroups. Newsgroups include announcements of new newsgroups, web sites, and mailing lists.

rec   Topics concerning recreational activities and hobbies such as fishing, automobile racing, and golf.

sci   Topics related to science. Newsgroups include cryptography, geology, environment, and engineering.

soc   Topics that address social issues or just socializing. Many of these newsgroups focus on the cultures of particular countries.

talk   Newsgroups that generally debate topics such as abortion, gun control, politics, and religion. Discussions are lively but generally do not contain much useful information (of course there are always exceptions).

Other, smaller categories include:

ba   Newsgroups for discussion of issues in the San Francisco Bay Area.

biz   Topics related to business.

ca   Newsgroups for discussion of California issues.

Most newsgroups are unmoderated, which means that anybody can post a message about anything. Because this sometimes leads to excessive and useless postings, moderated newsgroups were created. In a moderated newsgroup, your message is sent to someone who looks at it and decides whether to allow it to be posted to the newsgroup. These newsgroups generally have a lot more useful information than similar unmoderated newsgroups.

➾ 2.7.1 Finding Useful Newsgroups

Similar to search engines that catalog web pages, there are sites on the Internet that contain catalogs of newsgroups.

One very useful engine for searching through most newsgroups is available at:
http://www.dejanews.com

For example, if you wanted to find a newsgroup that discussed bioremediation, you would use your web browser to go to the site indicated above. You would then follow the prompts to request a search for bioremediation. The search engine then returns a list of messages regarding bioremediation. You can read the messages to determine who posted the message and to which newsgroup the message was posted, and thus determine newsgroups that focus on topics of particular interest to you.

Figure 2.7.1A shows the DejaNews home page. Figure 2.7.1B shows a DejaNews search form.

SEARCH   HELP   CONTACTS   CREDITS

"Put simply, Usenet is the largest information utility in existence"
-Harley Hahn, *"A Student's Guide to UNIX"*

DejaNews is *the* tool for searching Usenet!

 It's big!
The largest collection of indexed archived Usenet news anywhere!

 It's fast!
Searches through mountains of Usenet archives in seconds to find the information you need.

 It gets results!
Fill-out forms and "how-to" guides help you target your search to get what you want, fast and easy.

 It's what you need!
The information is out there — now there's a way to find it! A must for your hotlist!

Figure 2.7.1A

DejaNews query form

search for :

───

options

☐ Maximum number of hits : 30 60 120
☐ Hitlist format : terse verbose
☐ Sort by (per page) : score group date author
☐ Default operator : OR AND

create a query filter to limit your search by newsgroup, date, or author

☐ **Age Bias**

    prefer new prefer old
    age matters: not at all somewhat greatly

───

power users' options

☐ (coming soon)

Figure 2.7.1B

2.8 BULLETIN BOARDS

Bulletin board systems (BBSs) are computer systems that you can connect to by dialing directly into the computer system using a terminal program. Some of these systems and the information contained on them are not available via the Internet.

These computers are often not part of the Internet or are only connected to the Internet for a limited time each day. However, there are many BBSs that contain very useful information. Consequently, if the computer that contains the BBS you want to use is a long way away, you will have to call long distance to connect to it, instead of making a local call to your local ISP.

2.8.1 How to Find Bulletin Board Systems

To our knowledge, there are no catalogs of BBSs available at the present time. The easiest way to find a relevant BBS is to post a question to a mailing list or newsgroup. Others who have been there before are generally very helpful in letting you know where you can get BBS information.

2.8.2 Accessing Bulletin Board Systems

For your computer, you need a terminal program, which likely came with your modem when you purchased it. Programs such as *Procomm Plus* and *Smartcom* for PCs and *Zterm* for Macs can be used. These terminal programs can allow your computer to emulate other terminals. Standard terminal emulations include VT100 and ANSI. It is not really important what these emulations mean, but it is important to make sure that your terminal is set to an emulation that the bulletin board understands.

Where possible, we have given you the terminal setup settings you need to access the BBSs. Go to your terminal program, go to the terminal settings selection on the menu, and set your terminal accordingly.

Then all you need to do is dial the bulletin board computer's phone number and follow instructions.

3

Environmental Information on the Internet for Free

The most important thing to remember when retrieving information using the Internet is: **Let the browser (surfer) beware**. *Because the Internet is growing so rapidly, information quickly becomes obsolete. Some pages may move or cease to exist. Some sites are better maintained than others. The information available via the Internet may not, in some cases, be the most current information available, as old habits of printing onto paper die hard.*

3.1 REGULATORY INFORMATION

This section describes regulatory and legal information that can be accessed via the Internet. It is organized to cover first international environmental standards, then federal regulations, and then California state standards, as an example of the state information that is or may in the future be on the Internet for your state.

3.1.1 International Standards

The International Organization for Standardization (ISO) is in the process of developing international environmental standards, known collectively as ISO 14000. Information regarding these standards can be found in a number of newsgroups. The URLs for the most commonly used newsgroups are provided below:

news:sci.environment
news:sci.engr.safety
news:misc.industry.quality

There is an unmoderated Majordomo mailing list designed for the discussion of the ISO 14000 certification guidelines for environmental and related industries.

> To subscribe, send an email message to *majordomo@quality.org.*
> The message body should contain the following as the only text:
> > **subscribe iso14000**
> The List Administrator is Gene Tatsch *CET%LOUIE@rcc.rti.org.*

Another ListServer mailing list is also devoted to the discussion of ISO 14000 issues.

> To subscribe, send an email message to *Listserv@vm1.Nodak.edu*
> The message body should contain the following as the only text:
> > **subscribe QUEST firstname lastname**

3.1.2 Federal Regulations

This section describes where you can find the Code of Federal Regulations (CFR), Federal Register, federal bills, the history of bills and resolutions, the congressional record and its index, and the U.S. Code online.

Often the information exists at more than one site and can be accessed in more than one way. For example, you can access GPO databases through the GPO home page on the web, or you can telnet to the databases that are maintained at various libraries throughout the U.S. Some of the information can be accessed through gopherspace and other parts of the databases can be sent to you automatically through email.

Finally, this section describes a number of commands that can commonly be used to search regulatory databases.

3.1.2.1 The Code of Federal Regulations

The complete text of the CFR is available on the web at:
http://www.pls.com:8001/his/cfr.html

It can also be reached using gopher, but no search ability is provided at the gopher site. Note that this is a demonstration site; the CFR that is posted is the 1992 CFR. An updated CFR should become available from the House Information Systems directly (rather than through PLS) in January 1996.

Figure 3.1.2.1 shows the U.S. House of Representatives Internet Law Library Code of Federal Regulations page.

Online help for using the CFR and U.S. Code databases is available at:
http://www.pls.com/plweb/info/help/oltoc.html

3.1.2.2 The Federal Register

The FR is available online, for free, through the web, through telnet, and parts of it are available through email subscription services.

The GPO provides searchable web access to the following databases (and other databases described in the next section):

The 1996 FR (vol. 61, number 1) is updated by 6:00 a.m., Eastern time, each day that it is published.

The 1994 and 1995 FR databases contain the daily issues of the FR from vol. 59 and vol. 60. All information in the FR is included.

The 1995 and 1996 Unified Agenda, otherwise known as the Semi-Annual Regulatory Agenda, which summarizes pending regulatory actions by federal agencies, is also available. The database is updated semi-annually (usually April and October) when the Unified Agenda is published. The 1994 Unified Agenda is also available.

The U.S. House of Representatives Internet Law Library Code of Federal Regulations

Welcome To the Code of Federal Regulations!

We invite you to discover the Code of Federal Regulations experimental server. This experimental server is brought to you by House Information Systems in cooperation with Personal Library Software, Inc.

This server is experimental, to demonstrate the potential for making the CFR available to the public at no charge. If you are using it for legal research, we urge you to verify your results with the printed Code of Federal Regulations available through the U.S. Government Printing Office.

(1) Describe what you are looking for. _____

(2) Select an operation. Click here for a description of search or advisor functions.
Search Concept Search —— **Advisors:** Relate — Dictionary — Fuzzy

(3) Click here to

Your Comments Please!
Let us know if you have any comments about the presentation of this information, or any ideas or concerns you want to convey to the House of Representatives. (Comments@hr.house.gov)

Figure 3.1.2.1

The GPO provides web access to the FR and Unified Agenda databases described above at the following sites:

http://www.okstate.edu/gpolink.html — Oklahoma State University
http://thorplus.lib.purdue.edu/gpo — Purdue University
http://ssdc.ucsd.edu/gpo — University of California, San Diego
http://www.lib.utk.edu/gpo/GPOsearch.html — University of Tennessee

Each of these sites is slightly different. Some will let you search multiple databases, while others will require that you search one at a time. Otherwise, the content and search mechanisms operate the same.

☞ Some searches may return a *.pdf* or *.tiff* file in addition to ASCII text. This means that with the right software, you can print out the documents exactly as they appear in the FR.

☞ For those looking for more technical information, a *.pdf* file indicates that the document was created by Adobe Acrobat and can therefore be read by the Adobe Acrobat reader. This reader is available at *http://www.adobe.com.*

☞ The great thing about *.pdf* is that you can download the reader for free and then see and print out the documents exactly as they appear in the Congressional Record or FR.

☞ *.tiff* files are graphics that may be viewed and printed using graphics programs such as PageMaker, PhotoStyler, PhotoPaint, etc.

Users can also access these databases by telnetting to the following sites at which GPO databases are maintained:

Telnet to:	*sled.alaska.edu*	Telnet to:	*LIAS.psu.edu*
User ID:	**Sled**	User ID:	None
Password:	**Sled**	Password:	None
Telnet to:	*aztec.asu.edu*	Telnet to:	*lib.ursinus.edu*
User ID:	**Guest**	User ID:	**Library**
Password:	**Visitor**	Password:	None

Telnet to:	*pldvax.pueblo.lib.co.us*	Telnet to:	*osfn.rhilinet.gov*
User ID:	**Library**	User ID:	**Visitor**
Password:	None	Password:	None
Telnet to:	*library.unc.edu*	Telnet to:	*atlas.tntech.edu*
User ID:	**Library**	User ID:	**PAC**
Password:	None	Password:	None
Telnet to:	*catalog.cwru.edu*	Telnet to:	*msuvx1.memphis.edu*
User ID:	None	User ID:	**Library**
Password:	None	Password:	None
Telnet to:	*portals.pdx.edu*	Telnet to:	*link.tsl.texas.gov*
User ID:	None	User ID:	**Link**
Password:	None	Password:	None
Telnet to:	*spl.lib.wa.us*		
User ID:	**Library**		
Password:	None		

For example, detailed instructions for accessing the Seattle Public Library database are provided below:

1. Telnet to: *198.137.188.2* (or *spl.lib.wa.us*)
2. Login as: "library"
   [There are several terminal emulation questions.
   Most people use vt100]
3. Select: GPO Access (enter the number)
4. Select: GPO Access (enter the number)
5. Enter your terminal setting.
6. At this point, you will be provided with a numbered list of databases which you can access. To select a database:
   a. use the arrow keys to highlight a database, then
   b. use the space bar to asterisk that choice, then
   c. hit enter.
7. You will be prompted to enter keywords to retrieve an index of documents to your screen. [If you are looking for a particular bill, i.e., S. 540, it is useful to use the syntax "s. ADJ 540" (all boolean operators, such as ADJ (adjacent to), must be in all capitals).]

8. The system will return an index of all of the documents that contain the keywords. Enter the number of the document that you want to look at and hit enter.
9. To save a document, go back to the index by typing "q" and then:
    a. highlight the document by using the arrow keys or typing the document's number (the left most number on the screen), then
    b. enter "m" (to mail it), then
    c. enter your entire email address.
10. You must logoff the system before hanging up. To do this:
    a. select "q", then
    b. select "3" Logoff.

The U.S. Environmental Protection Agency (EPA) also has a ListServer that distributes selected FR documents automatically via email on the day of publication.

☞ Both ASCII files and the corresponding *.tiff* graphics will also be accessible via the EPA public access gopher (*gopher.epa.gov*). The ListServes are described as follows:

LISTSERVE NAME	DESCRIPTION
EPAFR-CONTENTS	Contains the full-text FR Table of Contents with page number citations.
EPA-MEETINGS	Contains all meeting notices. Program-specific meetings would be duplicated under that program.
EPA-SAB	Material relating to the Science Advisory Board.
EPA-IMPACT	All environmental impact statements published in the FR.
EPA-SPECIES	All endangered species documents published in the FR.
EPA-GENERAL	All general EPA non-program-specific documents, presidential documents related to environmental issues, and other agency documents other than environmental impact and endangered species actions.

EPA-AIR	All Office of Air and Radiation documents.
EPA-PEST	All Office of Pesticide Programs documents.
EPA-TOX	Office of Pollution Prevention and Toxic Substances documents excluding Community-Right-To-Know (Toxic Release Inventory) documents.
EPA-TRI	Community-Right-To-Know Toxic Release Inventory documents.
EPA-WASTE	All Hazardous and Solid Waste documents.
EPA-WATER	All Office of Water documents.

To subscribe to any of the above ListServes, send an email message addressed to: *Listserver@unixmail.rtpnc.epa.gov.*

The body of the message should read Subscribe <listname> <your first name> <your last name>. For example: **Subscribe EPA-TRI Dorothy Thornton.**

3.1.2.3 The U.S. Code, Bills, and Congressional Record

The complete text of the U.S. Code, Bills, and Congressional Record is available through the web, gopher, and telnet.

The GPO provides searchable web and telnet access to the following seven databases (and other databases described in the previous section):

Bills. All published versions of bills from the 104th (1995-1996) and 103rd Congress (1993-1994). Bills from the 104th Congress are updated by 6 a.m. daily when bills are published and approved for release.

History of Bills and Resolutions. The 1995 History of Bills and Resolutions is a section of the 1995 Congressional Record Index (vol. 141) that provides information about all bills and resolutions introduced during the 1st Session of the 104th Congress. The database is updated biweekly when the Congressional Record Index is published.

The 1994 History of Bills and Resolutions database provides information about all bills that were introduced during the 2d session or introduced during the 1st session and acted upon during the 2d session of the 103d Congress. Entries for each bill include actions that are reported in the Congressional Record and reference the issue date and pages where the action is reported. Because biweekly issues of the History of Bills do not cumulate, there is no equivalent print counterpart of these databases.

The Congressional Record. The Congressional Record for the 104th Congress, 1st Session from January 4, 1995 (vol. 141, number 1). It is updated between 9:00 a.m. and 11:00 a.m. Eastern time each day the Congressional Record is published, depending on when Congress adjourned the previous day. The daily issues of the Congressional Record for the 103d Congress, 2d Session (1994) are also available (vol. 140).

The Congressional Record Index. The 1995 Congressional Record Index (vol. 141) provides information about the proceedings and debates of the U.S. House of Representatives and U.S. Senate during the 104th Congress, 1st Session, as published in the Congressional Record.

The database is updated biweekly when the Congressional Record Index is published. Entries include actions that are reported in the Congressional Record and reference the issue date and pages where the action is reported. The 1992, 1993 and 1994 Congressional Record Index databases cover the 102d Congress, 2d Session, and the 103d Congress, 1st and 2nd Sessions, respectively. Because biweekly issues of the Congressional Record Index do not cumulate, there is no equivalent print counterpart of these databases.

The U.S. Code. The U.S. Code is prepared and published by the Office of the Law Revision Counsel, U.S. House of Representatives, and contains the general and permanent laws of the U.S. in effect as of January 1994 or January 1995, depending on the title. The database is updated periodically, annotating changes to individual sections. Whole titles are superseded annually as the editing process for each title is completed. Notification of a change to a section of the U.S. Code usually occurs within five business days of the enactment of the law; the text of a revision is incorporated when the annual bound volumes are published.

Public Laws. The Public Laws database is a collection of laws enacted during the 104th Congress (1995-1996). The header of each section indicates whether a recent Public Law affects that particular section, but the text remains unchanged until the annual revision cycle. Prepared and published by the Office of the Federal Register (OFR), National Archives and Records Administration, each law is first published as a slip law and then later compiled into a volume of the Statutes at Large. The Public Laws database contains the text of each law enacted and is updated irregularly as the publication of a slip law is authorized by the OFR.

The Government Accounting Office (GAO) Blue Book Reports. GAO "blue book" reports from October 1, 1994. The database is updated daily and reports are available within two business days after public release. The reports contain GAO's findings and recommendations to members of Congress or to congressional committees. Restricted and classified products, as well as correspondence, are not included in this database.

These seven GPO databases described above can be accessed at the same sites as the GPO databases described in the previous section. The GPO database web addresses are repeated below for your convenience:

> *http://www.okstate.edu/gpolink.html* — Oklahoma State University
> *http://thorplus.lib.purdue.edu/gpo* — Purdue University
> *http://ssdc.ucsd.edu/gpo* — University of California, San Diego
> *http://www.lib.utk.edu/gpo/GPOsearch.html* — University of Tennessee

The U.S. Code can also be accessed at the following web site:
> *http://www.pls.com:8001/his/cfr.html*

The U.S. Code can also be accessed at the following gopher site:
> *gopher://hamilton1.house.gov:70/1*

You cannot search through the code for a specific reference at this site.

Free 24-hour online access to the text of U.S. Senate and House bills and the Congressional Record is available through the U.S. GPO. The telnet addresses are provided in section 3.1.2.2 above.

3.1.3 State Regulations—California

California air pollution statutes and regulations. Online information includes the District Rules Database which contains the Current Rule Books for 29 California air districts, available at:

http://arbis.arb.ca.gov/html.

California Water Code and Regulations. The California Water Code and programs can be reached through the State Water Resources Control Board (SWRCB) home page. The regulatory information is located at:

http://www.swrcb.ca.gov/pub/

Recently adopted water quality plans and standards and recently adopted water rights orders and decisions are also available at this site, as is the SWRCB Chapter 15 Guidance and Policies.

These files can also be accessed via the SWRCB Electronic Bulletin Board System. The telephone number for the BBS is (916) 657-7922 (916) OK-SWRCB). However, the modem speed for the BBS is 2400 baud, which makes it very, very slow.

California Hazardous Waste Code and Regulations. These can be reached through the Department of Toxic Substances Control (DTSC) web page. A list of available documents can be found at the following address:

http://www.cahwnet.gov/epa/dtsc.htm#TAG4

Documents include an introduction to California hazardous waste laws, RCRA authorization regulations, AB 1772-Tiered Permitting, permit by rule for fixed treatment units, permit by rule for transportable treatment units, hazardous waste codes, a summary of bills through October 1994, and the 1993-94 Summary of Legislation.

3.1.4 Searching Regulatory Databases

Most regulatory databases use what is known as natural language to search for information. Just typing in what you are looking for should be relatively effective. The search engine generally uses the following operators to help you tailor your search for information:

ADJ	Returns records that contain the second word immediately after the first word. Ex: **hazardous ADJ waste**.
AND	Returns records that contain both of the words (or expressions). For example, if you want to know if there are any hazardous waste regulations that pertain to solvents, you might enter the following search request: Ex: **(hazardous ADJ waste) AND solvent***.
OR	Returns records that contain either of the words (or expressions). Ex: **(storm ADJ water) OR (stormwater)**.
NOT	Returns records that contain the first word (or expression) but not the second. For example, if you want to search for regulations regarding hazardous materials, but are not interested in knowing anything about hazardous waste regulations, you might search for **hazardous NOT waste**.
NEAR	Returns records that contain the words (or expressions) near each other. For example, the following search statement would return records containing "stormwater" within 3 words of "industrial" (articles and conjunctions like "a", "and", and "the" are not counted): **stormwater NEAR/3 industrial**.
W/n	Returns records that contain the second word within **n** words after the first word. For example, if you want to find out if there are training requirements described for hazardous materials or hazardous waste workers, you might start by searching for the word training within 50 words of hazardous: **hazardous W/50 training**.
?	Is a single character wildcard. For example, entering **260.1?** will return 260.10 through 260.19, as well as 260.1a through 260.1z.
*	Is a multiple character wildcard used with the root of a word that might have a number of different suffixes. For example, **solvent*** will return solvent and solvents, and **train*** will return train, trains, training, trained, trainer, trainee, etc.

\#   Forces an exact match. The following would find start, but not starter, starting, or started: **start#**.

@   Allows similar words to be used to match the word. The following would find car and automobile: **auto@**.

The operators such as AND, OR, NOT, and ADJ must be entered in capital letters.

The CFR databases use four fields to identify each part of the CFR:

CITE   EXPCITE
HEAD   TEXT

For example, the CITE field for Title 40 of the CFR Part 260 Section would contain "40 CFR Sec.260.10".

So if you wanted to search for this specific reference, you would type:
CITE=40 ADJ CFR ADJ Sec.260.10

If you wanted to find the OSHA Asbestos Standard which you knew was located at 29 CFR 1910.1001, you would type:
CITE=29 ADJ CFR ADJ Sec.1910.1001 or
CITE=29 ADJ CFR ADJ Sec. ADJ 1910.1001

Note that the difference is simply a space between "Sec." and "1910.1001".

3.2 CHEMICAL-SPECIFIC INFORMATION

This section describes a number of web sites at which chemical-specific information can be found. The following sites are described in this section:

1.   EPA's Chemical Substances Database

2.   The Agency for Toxic Substances Research's (ATSDR's) Hazdat database

3. The University of Utah's Material Safety Data Sheet (MSDS) database

4. The NTP's Abstracts of Toxicological Studies Database

5. The Right-to-Know Computer Network's (RTK NET's) New Jersey Fact Sheets

6. Envirofacts Master Chemical Integrator (EMCI)

7. Water Resources Scientific Information Center (WRSIC) of the U.S. Geological Survey (USGS) Selected Water Resources Abstracts database

8. California's DTSC Information

9. California's Office of Environmental Health Hazard Assessment (OEHHA) Proposition 65 list

We have included a number of samples of printouts from the sources described below so that you can evaluate the depth of information provided. These samples are attached as Appendix A.

3.2.1 EPA's Chemical Substances Database
gopher://ecosys.drdr.Virginia.edu:70/11/library/gen/toxics

This database is maintained by the University of Virginia and contains information regarding hundreds of chemicals. The date the information was compiled is listed at the beginning of each record. The record for each chemical in the database includes:

- the common name
- CAS number
- DOT number
- a hazard summary
- identification information (color, odor, viscosity, etc.)
- the reason the chemical is cited in the database
- how a person can determine if he or she is being exposed to the chemical
- workplace exposure limits (note the date the record was prepared, these limits might have changed)
- ways of reducing exposure

Figure 3.2.1 in Appendix A shows a copy of the report for acetone.

3.2.2 Hazdat
http://atsdr1.atsdr.cdc.gov:8080/hazdat.html

Hazdat is ATSDR's Hazardous Substance Release/Health Effects Database. Figure 3.2.2A shows the Hazdat's index of chemical files page (its ToxFAQs menu).

Hazdat contains substance-specific health effects information for more than 150 hazardous substances. The ATSDR Public Health Statement for acetone is included as Figure 3.2.2B in Appendix A.

Internet HazDat may also be queried directly from a web browser with Forms support by clicking on any one of the database queries listed at the Hazdat home page. Users can make site activity queries, site contaminant queries, toxicological profile queries, and public health statement text searches.

3.2.3 MSDSs

The University of Utah maintains a database of MSDSs for hundreds of chemicals. Access is presently (July 1996) free. The address of the site is:
    gopher://atlas.chem.utah.edu:70/11/MSDS

Figure 3.2.3 in Appendix A shows an MSDS for acetone obtained from the database.

3.2.4 NTP's Abstracts of Toxicological Studies
http://ntp-server.niehs.nih.gov/Main_Pages/NTP_STUDY_PG.html

The NTP is a part of the National Institutes of Health. It was established in a cooperative effort within the Public Health Service and the Department of Health and Human Services to:

- coordinate toxicology research and testing activities
- provide information about toxic chemicals
- strengthen the science base in toxicology

 ToxFAQs™

Agency for Toxic Substances and Disease Registry
Atlanta, GA

ATSDR ToxFAQs™ MENU

INDEX of CHEMICAL FILES

A B C D E F G H I J K L M N O P Q R S T U V W X Y Z

(For a Quick Search, Click on the first Letter of the Chemical / Hazardous Substance)

ATSDR ToxFAQs™ is a series of summaries about hazardous substances being developed by the ATSDR Division of Toxicology. Information for this series is excerpted from the ATSDR Toxicological Profiles and Public Health Statements. Each fact sheet serves as a quick and easy to understand guide. Answers are provided to the most frequently asked questions about exposure to hazardous substances found around hazardous waste sites and human health effects.

If you didn't find your chemical on the ATSDR ToxFAQs™ MENU, try searching the following files:

- ATSDR/EPA Top 20 Hazardous Substances
- ATSDR Public Health Statement Text Search
- Medical Management Guidelines for Acute Chemical Exposures: Chemical Protocols

---

INDEX of Chemical Files

- A -

☐ Acetone
☐ Aldrin/Dieldrin
☐ Aluminum
☐ Antimony
☐ Arsenic

Figure 3.2.2A

Online abstracts of NTP reports are available. The addresses of the study databases are listed below.

Long-Term Toxicological Studies
http://ntp-server.niehs.nih.gov/htdocs/pub.html

Short-Term Toxicological Studies
http://ntp-server.niehs.nih.gov/htdocs/pub-ST.html

Immunotoxicity Studies
http://ntp-server.niehs.nih.gov/htdocs/pub-IT.html

Reproductive Toxicology Studies
http://ntp-server.niehs.nih.gov/htdocs/pub-RT.html

Teratology Studies
http://ntp-server.niehs.nih.gov/htdocs/pub-TT.html

Examples of the information contained in these databases are provided in the following figures in Appendix A:

Long-Term Toxicological Studies   Figure 3.2.4A
Short-Term Toxicological Studies   Figure 3.2.4B
Immunotoxicity Studies   Figure 3.2.4C
Reproductive Toxicology Studies   Figure 3.2.4D
Teratology Studies   Figure 3.2.4E

3.2.5 RTK NET's New Jersey Fact Sheets
telnet://198.3.148.6:23

RTK NET is used mostly by environmental and public interest groups. It is designed to provide easy public access to EPA's Toxic Release Inventory (TRI) information.

The New Jersey Fact Sheets provide health effects information on each TRI chemical. The information in this database was last updated in 1991.

3.2.6 EMCI

http://www.epa.gov/enviro/html/emci/emci_overview.html

EMCI is used to integrate the varied chemical identifications used in four other EPA databases (PCS, RCRIS, CERCLIS, and TRIS described elsewhere in this manual).

EMCI provides users with a cross reference to chemical data reported in program office (for example, water versus air programs) systems. It uses an internal registry system based on Chemical Abstracts Service (CAS) Registry Numbers. The EMCI has particular benefits for the integration of chemicals where:

— Different program office systems identify the same chemical substance by different chemical names (i.e., synonyms) and different chemical identification codes (e.g., CAS Registry Numbers, PCS parameter codes, and RCRA hazardous waste codes).
— One program office system stores data for specific chemical substances (e.g., "1,1-dichloroethene" and "1,2-dichloroethene," both cis and trans) and other systems contain data for the non-specific category (e.g., "dichloroethene").

The integrator eliminates the need for a user to know how a chemical substance of interest is identified in the various systems when accessing environmental data. The integrator facilitates integrated access for the following types of chemical substances:

— Specific, unique chemical substances for which Chemical Abstracts Service Registry Numbers (CASRN) have been assigned.
— Regulated entities for which CASRNs have not been assigned (e.g., categories of chemically related substances, mixtures, and related substances with varying positional and spatial characteristics).

EMCI enables you to search for chemicals by CASRN or by name fragments for either common name or CAS index name. It identifies program office systems that contain data for a particular chemical substance or category of substances. Only a small number of common names are available in EMCI at this time.

Chemical groupings (i.e., components of categories, mixtures, and non-specific substances) were captured from the EPA Register of Lists or were constructed manually by EMCI data administrators, based on database searches. Cross references for substance relationships are still incomplete in this version of EMCI.

3.2.7 USGS's WRSIC

http://www.uwm.siu.edu/databases/wrsic/

This site contains the Selected Water Resources Abstracts database. The database is an international collection of water research compiled by the WRSIC. Research selected to be abstracted includes research in the life, physical, and social sciences, as well as engineering and legal issues. This database covers the time period from 1967 to October 1993.

The database contains over 265,000 abstracts and citations. Each citation gives full bibliographic information so that you can then locate your titles of interest. Most can be found through normal library resources, using the information provided in the citation. Searching through this database is done through user-provided keywords. Given the tremendous size of this database, a carefully chosen set of keywords is needed to locate the types of abstracts you desire.

A list of the general chemical and health areas covered by the database is provided below:

- chemistry, toxicology
- public health, hygiene, biomedical

3.2.8 California's DTSC Information

California's DTSC has a large number of chemical- or mixture-specific documents available at:

http://www.cahwnet.gov/epa/dtsc.htm#TAG4.

These documents provide information about solvents, used batteries, lead in the construction industry, pesticides, fluorescent light bulbs, used oil, asbestos, PCBs, CFCs, and antifreeze, to name but a few. The information concentrates on how to manage the waste materials. Figure 3.2.8 in Appendix A provides an example of the information available through DTSC.

3.2.9 California's OEHHA Information

California's OEHHA provides a list of all Proposition 65 chemicals (carcinogens or reproductive toxins) at its web page. The page can be reached through Cal EPA's home page at:

http://www.cahwnet.gov/epa/oehha.htm

3.3 WATER INFORMATION

There are a large number of sites on the Internet that can provide useful water information to environmental managers. This section describes databases that are available on the Internet or on BBSs.

California and the San Francisco Bay Area electronic information services are also described, for two reasons. First, the state and region are among the leaders in the computerized supply of information. And second, we wish to point out that useful information can be found at both a state and local level. Obviously, this manual could not provide information regarding environmental electronic resources available in every state and locality, but hopefully these examples provide you with some ideas about what kind of information your state and locality might have online now or in the future.

The following databases are discussed in this section:

1. EPA's Permit Compliance System (PCS), the agency's water permitting database

2. RTK NET's PCS database

3. EPA's Office of Research and Development (ORD) BBS

4. USGS's Water Use Data

5. USGS's National Water Conditions

6. National Oceanographic Data Center (NODC)

7. WRSIC of the USGS

8. EPA's Online Library System (OLS)

9. Environmental Guidance Documents

10. The California SWRCB

11. The San Francisco Bay Regional Water Quality Control Board (SFBRWQCB) BBS

12. Environmental Engineering Mailing List

13. Environmental Newsgroups Relating to Water

3.3.1 PCS
http://www.epa.gov/enviro/html/pcs/pcs_overview.html

PCS is a national system that contains National Pollutant Discharge Elimination System (NPDES) data and tracks permit issuance, permit limits, monitoring data, and other data pertaining to facilities regulated under the NPDES.

PCS records water-discharge permit data on more than 75,000 facilities nationwide. Forms to conduct searches to determine if a facility is in the database are presently provided. PCS contains general descriptive information on each permitted facility such as:

- facility name
- mailing address
- standard industrial classification

PCS also contains a huge amount of additional information described below. Access to this additional information requires special software and setup. The connection requirements can be obtained from:
http://www.epa.gov/enviro/html/faq.html

PCS tracks events relating to the issuance of a permit, from initial receipt of the application for a permit through actual permit issuance, including:

- date of the permit issuance
- date of permit expiration
- public notice date

PCS contains detailed information describing each outfall within a permitted facility and the discharge monitoring requirements associated with each, including:

- limit start and end dates that indicate when a facility may discharge
- monitoring locations
- parameters to be monitored
- the quantity and concentration limits for each pollutant
- the units in which pollutant concentrations are reported

PCS also tracks reported measurement values for effluent parameters including those that are violations of established limits of the permit. The database tracks information related to violations of compliance schedules.

PCS also contains information describing inspections that have been performed at a permitted facility, including:

- date of the inspection
- type of inspection
- who conducted the inspection
- detailed information about pretreatment information gathered as part of an audit

The database contains data related to enforcement actions including:

- the events in violation
- dates of occurrence
- type of enforcement action(s)
- dates action(s) were taken
- current status of each action
- data related to evidentiary hearings held when permittees wish to appeal or negotiate limits or compliance schedule requirements

3.3.2 RTK NET's PCS
telnet://198.3.148.6:23

RTK NET is used mostly by environmental and public interest groups. It is designed to provide easy public access to EPA's TRI information. EPA's PCS database is available through RTK NET on a pilot basis.

3.3.3 ORD BBS

The ORD BBS is a text-searchable database of all ORD publications since 1976, including titles and abstracts. Documents can be ordered online. Subjects covered include:

- water research
- biotechnology research
- other research

Several electronic conferences can be accessed regarding:

- biotechnology
- regional EPA operations
- water regulations
- methods of standardization
- ORD research paper bibliographic database (this is also available online)

The BBS can be reached at:

Telnet:	None at present
Voice Number:	513-569-7272
Data Number:	513-569-7610
Modem Setting:	8 data bits, N parity, 1 stop bit, full duplex

3.3.4 USGS's Water Use Data
http://edcwww.cr.usgs.gov/glis/hyper/guide/1_250_lulc

Data regarding water use is available from the USGS.

3.3.5 USGS's National Water Conditions Data
http://nwcwww.er.usgs.gov:8080/NWC/

Data regarding groundwater extremes, groundwater aquifers, stream water extremes, water quality, and selected reservoirs is available from the USGS.

Figure 3.3.5A in Appendix A shows some of the water quality data available for the Ohio River. Figure 3.3.5B in Appendix A shows depth to groundwater for a number of aquifers. Figure 3.3.5C shows a map indicating streamflow during April 1996.

3.3.6 The NODC
http://www.nodc.noaa.gov/

The NODC provides buoy information and information regarding marine toxic substances and pollutants.

Figure 3.3.6 shows the NODC home page.

3.3.7 WRSIC
http://www.uwin.siu.edu/databases/wrsic/

This site contains the Selected Water Resources Abstracts database. The database is an international collection of water research compiled by WRSIC. Research selected to be abstracted includes research in the life, physical, and social sciences, as well as engineering and legal issues. This database covers the time period from 1967 to October 1993.

The database contains over 265,000 abstracts and citations. Each citation gives full bibliographic information so that you can then locate your titles of interest. Most can be found through normal library resources using the information provided in the citation.

Searching through this database is done through user-provided keywords. Given the tremendous size of this database, a carefully chosen set of keywords is needed to locate the types of abstracts you desire.

National
Water
Conditions

UNITED STATES
Department of the Interior
Geological Survey

CANADA
Environment Canada
Climate Information Branch
Quebec Environment

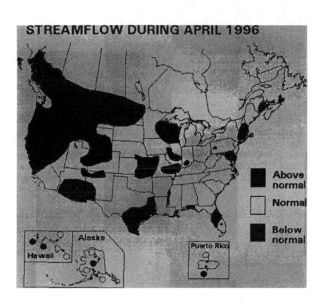

After viewing our information please use the NWC Comment Form to give us feedback!

[Forward to <u>Table of Contents</u>]
[Forward to the <u>USGS Water Resources Information Page</u>]

Figure 3.3.5C

 U.S. DEPARTMENT OF COMMERCE

 National Oceanic and Atmospheric Administration

National Oceanographic Data Center (NODC) Home Page

Welcome to the Mosaic server of the National Oceanographic Data Center. The NODC is one of the environmental data centers operated by the U.S. National Oceanic and Atmospheric Administration (NOAA). The NODC provides ocean data management and ocean data services to researchers and other users in the United States and around the world. For further information please click on a highlighted topic:

- What's New at NODC

- NODC Overview

- NODC Headquarters Offices in Silver Spring, Maryland

- NODC Liaison Offices

- How to Obtain NODC Products and Services

- How to Submit Data to NODC

- About Mosaic

- NODC's Gopher Server

- NODC Director Bruce Douglas goes to the University of Maryland

- NODC OCEAN BULLETIN BOARD SERVICE

Figure 3.3.6

A list of the general water areas covered by the database is provided below:

- municipal water supplies, water power
- water resources, planning and management
- aquaculture, fish and wildlife, zoology
- public utilities and policy, water law
- industrial water usage and treatment
- environmental and ecological data
- oceanic technology
- coastal, basin, and estuary data
- groundwater, water desalination
- fluid mechanics, physics

3.3.8 EPA's OLS
telnet://epaibm.rtpnc.epa.gov

EPA's OLS contains several databases including its Clean Lakes database which contains citations and summaries on topics relating to lake management, protection, and restoration. It can be reached at:

Telnet:	*epaibm.rtpnc.epa.gov*
Voice Number:	919-541-2777
Data Number:	919-549-0700 (2400)
	919-549-0720 (9600)
Modem Settings:	7 data bits, 1 stop bit, Even parity, Half duplex

3.3.9 Environmental Guidance Documents
http://www.tis.eh.due.gov/docs/egm/

The gopher site at the address given above provides information on the following water issues:

- the Clean Water Act (CWA)
- the Safe Drinking Water Act
- groundwater
- the Oil Pollution Act

Figure 3.3.9 in Appendix A contains a printout of the list of guidance documents available regarding the CWA and groundwater information.

3.3.10 The California SWRCB

The SWRCB has its own home page at *http://www.swrcb.ca.gov* and provides access to, among other things:

- SWRCB/RWQCB Addresses and Meetings Schedules
- Application Forms – Water Quality
- Application Forms – Water Rights
- Recent Bulletins and Announcements
- Fees
- Glossary and Cross Reference of Terms
- Minutes of Meetings
- Nonpoint Source Information
- Proposed Statewide Water Quality Plans
- Publications & Other Sources of Info
- Bay Delta Hearing Transcripts
- Staff Reports/Agenda Items
- All about the SWRCB Organization

The file areas described above can also be accessed via the SWRCB BBS. The telephone number for the BBS is (916) 657-7922 (916 OK-SWRCB).

Figure 3.3.10 shows the SWRCB home page.

3.3.11 The SFBRWQCB

The SFBRWQCB bulletin board provides a wealth of information about environmental programs and sites over which the Board has jurisdiction.

From the main menu, you can go to conferences to read what other people have written, ask questions, or send email to other users of the bulletin board. Available conferences include:

State Water Resources Control Board

Figure 3.3.10

- Pollution Prevention
- Surface Water
- Groundwater
- Nonpoint Source
- Basin Plan
- Department of Defense Site Cleanups

You may also use SFBRWQCB databases or download their database files. For example, you can download the SFBRWQB's entire database of leaking underground storage tank sites for the Bay Area or download the database for just one county. You can also download the list of North Bay toxic sites — those sites that have reported non-fuel releases.

The following databases are also available online and may be searched:

- NPDES and WDR permittees
- Leaking Underground Storage Tank System Information System (LUSTIS)
- Stormwater NOIs – Industrial
- Stormwater NOIs – Construction
- Toxics
- Landfills in the San Francisco Bay Area
- CWC 13271 report (spills)

Telnet: None at the present time
Data Number: (510) 286-0404
Modem Settings: 8 data bits, 1 stop bit, No parity, ANSI preferred
                   terminal setting

3.3.12 Environmental Engineering Mailing List

ENVENG-L is a mailing list for the discussion of environmental engineering practice, education, and research. Environmental engineering is broadly defined to include waste treatment and water supply.

To subscribe, send an email message to *listproc@pan.cedar.univie.ac.at*. The message body should contain the following as the only text:

> **subscribe enveng-l <your first name> <your last name>**
> For example: **subscribe enveng-l John Doe**

3.3.13 Environmental Newsgroups

Newsgroups are a great place to find current discussions in a particular field and to find FAQs and pointers to web sites, mailing lists, and other resources. However, most of the newsgroups are unmoderated, leading to a lot of noise (irrelevant messages). Some of the newsgroups to check out include:

news:sci.environment   *news:sci.engr*
news:sci.geo.hydrology   *news:sci.engr.civil*
news:sci.engr.chem

3.4 LAND/SOIL INFORMATION

There are a large number of sites on the Internet that can provide useful soil/ land information to environmental managers. This section describes databases that are available on the Internet or on BBSs.

California electronic information services are also described, for two reasons. First, the state is among the leaders in the computerized supply of information. And second, we wish to point out that useful information can be found at the state level. Obviously, this manual could not provide information regarding environmental electronic resources available in every state, but hopefully these examples provide you with some ideas about what kind of information your state might have online now or in the future.

The following databases are described in this section:

1. The USGS's Radon Database

2. WRSIC's Selected Water Resources Abstracts database

3. The USGS's California Earth Science Data

4. The USGS's Land Use and Land Cover Data

5. The National Oceanic and Atmospheric Administration's (NOAA's) National Geophysical Data Center (NGDC)

6. California's SWRCB Underground Storage Tank (UST) information

7. California's DTSC

8. Environmental Guidance Documents

9. The Environmental Engineering Mailing List

10. Land/Soil Environmental Newsgroups

3.4.1 USGS's Radon Database

USGS's radon homepage is located at:
http://sedwww.cr.usgs.gov:8080/radon/radonhome.html

The USGS has information on the radon potential of rocks, soils, and water for the U.S. as a whole, as well as more detailed radon risk assessments in specific geologic environments, and investigations of the correlations between geology and indoor radon. EPA radon publications are available at:
http://www.epa.gov/RadonPubs

Figure 3.4.1 shows a map of geologic radon potential in the U.S. that is available through the USGS radon home page.

3.4.2 WRSIC
http://www.uwin siu.edu/databases/wrsic/

This site contains the Selected Water Resources Abstracts database. The database is an international collection of water research compiled by WRSIC. Research selected to be abstracted includes research in the life, physical, and social sciences, as well as engineering and legal issues. This database covers the time period from 1967 to October 1993.

U.S. Geological Survey

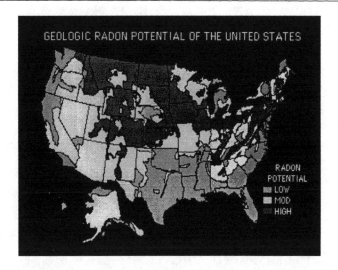

Click on the map to download a larger version (18kb)

The U.S. Geological Survey has compiled a series of geologic radon potential assessments for the United States in cooperation with the U.S. Environmental Protection Agency (EPA). The geologic radon potential reports describe the geology, soils, radioactivity, generalized housing construction characteristics, and other relevant information, and include discussions of the geologic factors controlling radon potential in each state. The geologic radon potential areas are defined by geologic boundaries, so that rock and soil units with similar radon generation and transport characteristics are grouped and delineated in the reports.

HOW WE DID THE ASSESSMENT: Methods and data used

The USGS geologic radon potential books are organized by EPA region, with individual chapters for each of the 50 states. The reports also include summaries of existing indoor radon data for each state.

Return to USGS Radon Home Page

URL: http://sedwww.cr.usgs.gov:8080/radon/rnus.html
Contacts: Randy Schumann, rschumann@usgs.gov or Linda Gundersen, lgundersen@usgs.gov

Figure 3.4.1

The database contains over 265,000 abstracts and citations. Each citation gives full bibliographic information so that you can then locate your titles of interest. Most can be found through normal library resources using the information provided in the citation.

Searching through this database is done through user-provided keywords. Given the tremendous size of this database, a carefully chosen set of keywords is needed to locate the types of abstracts you desire.

A list of the general land/soil areas covered by the database is provided below:

- agriculture, irrigation and drainage
- environmental and ecological data
- botany and soil science research
- forestry, alpine, glaciology, limnology
- civil engineering and transportation
- finance and economics, consulting
- horticulture, agronomy
- petrology, commercial resources

3.4.3 USGS's California Earth Sciences Data
http://wrgis.wr.usgs.gov/docs/geologic/ca/california.html

The USGS's Western Region Geologic Information Center contains California data sets, as well as data regarding other western states. Data is stored in *.tar* files and thus needs to be viewed in a Unix system. If you have a shell account, you can usually use the ISP computer to open the tar file and read the data.

3.4.4 USGS's Land Use and Land Cover Data
http://edcwww.cr.usgs.gov/glis/hyper/guide/1_250_lulc

Data regarding land use and land cover is available from the USGS. Figure 3.4.4 shows the first screen of the relevant web page.

3.4.5 The NGDC
http://www.ngdc.noaa.gov/

The NGDC collects geophysical data. Figure 3.4.5 shows the NGDC home page.

USGS Land Use and Land Cover Data

Table of Content

Background

The Land Use and Land Cover (LULC) data files describe the vegetation, water, natural surface, and cultural features on the land surface. The United States Geological Survey (USGS) provides these data sets and associated maps as a part of its National Mapping Program. The LULC mapping program is designed so that standard topographic maps of a scale of 1:250,000 can be used for compilation and organization of the land use and land cover data. In some cases, such as Hawaii, 1:100,000 scale maps are also used.

Land Use and Land Cover Example (6.8 kb)

Compilation is based upon a classification scheme identified in the Appendix.

Extent of Coverage

Land Use and Land Cover (LULC) data are available for most of the contiguous United States and Hawaii.

Acquisition

Processing Steps

Manual interpretation of aerial photographs acquired from NASA high-altitude missions and other sources were first used to compile the land use land cover maps. Secondary sources from earlier land use maps and field surveys were also incorporated into the LULC maps as needed. At a later time, the LULC maps were digitized to create a national digital LULC database. The evolution of this process resulted in the creation of the Geographic Information Retrieval Analysis System (GIRAS).

Figure 3.4.4

National Geophysical Data Center

The National Geophysical Data Center (NGDC) manages environmental data in the fields of solar-terrestrial physics, solid earth geophysics, marine geology and geophysics, paleoclimatology, and glaciology (snow and ice). In each of these fields it also operates a **World Data Center** (WDC A) discipline center.

Data, meta-data, and information at NGDC are available via the Internet using Web browsers, Gopher, and Anonymous-FTP. While not all of our data holdings are available through NGDC's Geophysical On-Line Data (GOLD), new data and information are continually being added.

What's  at NGDC   Search NGDC's

More information is available for the following areas:

Solid Earth Geophysics;   Solar-Terrestrial Physics; Marine Geology & Geophysics

Paleoclimate Program;   DMSP Satellite Data Archive;   The National Snow and Ice Data Center

NGDC provides support to other national and international programs. More information is available on NGDC's support to coastal programs.

Forms-based searches of the **NOAA Data Set Catalog** are offered by **NOAA's Environmental Information Services**. An alternative telnet search is available.

Related services on the Internet from NOAA and other sources.

More about NGDC people and services

Figure 3.4.5

3.4.6 California's SWRCB

The SWRCB has its own home page at *http://www.swrcb.ca.gov* and provides access to:

- UST regulations
- UST Trust Fund information
- UST upgrade information

The file areas described above can also be accessed via the SWRCB BBS. The telephone number for the BBS is (916) 657-7922 (916) OK-SWRCB).

3.4.7 California's DTSC
http://www.cahwnet.gov/epa/dtsc

California's DTSC is the agency that regulates hazardous materials in the state of California. Among the documents regarding hazardous wastes that can be found at the DTSC web site are:

- Hazardous Materials Storage Tanks
- Underground Storage Tank Closure Study Report
- Portable Tanks/Secondary Containment

3.4.8 Environmental Guidance Documents
http://www.tis.eh.doe.gov/docs/egm

The gopher site at the address given above provides guidance information regarding radiation protection and the Oil Pollution Act.

3.4.9 Environmental Engineering Mailing List

ENVENG-L is a mailing list for the discussion of environmental engineering practice, education, and research. Environmental engineering is broadly defined to include waste treatment, water supply, air pollution control, solid waste management, noise and radiation protection, waste minimization, and related issues.

To subscribe, send an email message to *listproc@pan.cedar.univie.ac.at*. The message body should contain the following as the only text:
   subscribe enveng-l <your first name> <your last name>

3.4.10 Environmental Newsgroups

Newsgroups are a great place to find current discussions in a particular field and to find FAQs and pointers to web sites, mailing lists, and other resources. However, most of the newsgroups are unmoderated, leading to a lot of noise (irrelevant messages). Some of the newsgroups to check out include:

news:sci.environment   *news:sci.engr*
news:sci.geo   *news:sci.geo.geology*
news:sci.geo.petroleum   *news:sci.engr.civil*

3.5 AIR INFORMATION

There are a large number of sites on the Internet that can provide useful air information to environmental managers. This section describes databases that are available on the Internet or on BBSs.

California electronic information services have also been described for two reasons. First, the state is among the leaders in the computerized supply of information. And second, we wish to point out that useful information can be found at the state level. Obviously, this manual could not provide information regarding environmental electronic resources available in every state, but hopefully these examples provide you with some ideas about what kind of information your state might have online now or in the future.

The following databases are described in this section:

1. Office of Air Quality Planning's (OAQP's) Technology Transfer Network (TTN)

2. OAQP's Applied Modeling Research (AMR) BBS

3. Aerometric Information Retrieval System (AIRS) Executive USA

4. AIRS Executive International

5. California's Air Resources Board Information System (ARBIS)

6. WRSIC

7. NOAA's Data Set Catalog

8. NOAA's National Climatic Data Center (NCDC)

9. Environmental Guidance Documents

10. Environmental Engineering Mailing List

11. Environmental Newsgroups

3.5.1 OAQP's TTN

OAQP's TTN is a network of BBSs developed to provide information exchange in different areas of air pollution control, ranging from emission test methods to regulatory air pollution models.

Eighteen BBSs are currently available or planned on the network. They are described below.

EMTIC   Emission Measurement Technical Information Center: Provides access to emission test methods and testing information for the development and enforcement of national, state, and local emission prevention and control programs.

AMTIC   Ambient Monitoring Technology Information Center: Provides information and all federal regulations pertaining to ambient monitoring.

AIRS   Aerometric Information Retrieval System: Provides information and documentation on the use and acquisition of air quality and emissions data from the AIRS mainframe computer system.

BLIS   BLIS is a compilation of air permits from local, state, and regional air pollution control agencies.

NATICH   National Air Toxics Information Clearinghouse: Contains information submitted by EPA, state, and local agencies regarding their air toxics programs to facilitate the exchange of information among government agencies.

COMPLI   Stationary Source Compliance: Provides stationary source and asbestos compliance policy and guidance information.

NSR   New Source Review: Offers guidance and technical information within the NSR permitting community.

SCRAM   Support Center for Regulatory Air Models: Provides regulatory air quality models, meteorological data, documentation, as well as modeling guidance.

CHIEF   Clearinghouse for Inventories/Emission Factors: Contains the latest information on air emission inventories and emission factors. It provides access to tools for estimating emission inventories for both criteria and toxic pollutants.

CAAA   Clean Air Act Amendments: Provides information on the 1990 amendments, regulatory requirements, implementation programs, criteria pollutants, and technical analysis.

APTI   Air Pollution Training Institute: Describes current course offerings.

CTC   Control Technology Center: Offers free engineering assistance, a hotline, and technical guidance to state and local air pollution control agencies.

USC   User Support Center: Provides support for users by offering information on modems, downloading communication software, and other communication issues. Also provides public message area for users.

ORIA   Office of Radiation and Indoor Air: Disseminates information to promote actions to reduce exposure to harmful levels of radiation and indoor air pollutants.

USCAN   US/Canada Air Quality Agreement: Provides for the exchange of permitting information between the states along the border between the U.S. and Canada.

OMS   Office of Mobile Sources: Provides information pertaining to mobile source emissions, including regulations, test results, models, and guidance.

AIRISC Air Risc: Provides technical assistance and information in areas of health, risk, and exposure assessment for toxic and criteria pollutants.

SBAP   Small Business Assistance Program: Provides support to state and local small business assistance programs by serving as a communications network to share materials as well as new federal rules that have been developed relating to small business issues.

OAQP's TTN can be reached at:

Telnet:	*ttnbbs.rtpnc.epa.gov*
Voice Number:	919-541-5384
Data Number:	919-541-5742 (14,400)
Modem Settings:	8 data bits, No parity, 1 stop bit, terminal emulation VT100 or VT/ANSI, full duplex

3.5.2 OAQP's AMR BBS

The AMR BBS provides air quality atmospheric models. It can be reached at:

Voice Number:	919-541-1376
Data Number:	919-541-1325

3.5.3 AIRS Executive USA
http://www.epa.gov/docs/airs/aexec.html

AIRS Executive USA is a computer database that contains a select subset of data extracted from the AIRS database. Its user-friendly format guides you to air pollution information on U.S. ambient air, plant emissions sources, and mobile sources. AIRS Executive operates entirely in graphics mode, allowing you to view and print data in a variety of formatted maps, charts, and reports.

AIRS Executive is designed to provide quick access to a subset of air pollution information. Managers, supervisors, and non-technical staff will find it especially helpful for obtaining easy-to-read summaries of air pollution data.

AIRS Executive is a menu-driven program, in which options appear in the form of icons. From any menu, you can use your mouse or keyboard to select an option. The first menu contains icons which represent the type of information available in AIRS Executive.

The "information" section contains general information about AIRS and each of its five subsystems, as well as information about AIRS Executive. The "data" section has reports and charts that summarize air pollution values reported by monitoring sites and emissions estimates from major sources of pollution. For each report or chart you choose a geographic area, pollutant, and year; AIRS Executive displays data that meet your criteria. Several types of reports are available. The "contacts" section lists names and phone numbers of people to contact for further assistance on each AIRS subsystem.

The databases used by AIRS Executive to generate the reports and charts in the "data section" are created by extracting the data from the AIRS database on EPA's mainframe computer. AIRS Executive is updated monthly to keep up-to-date with changes in the AIRS database. Updates are published on the 10th day of each month.

AIRS Executive is not intended to replace the AIRS mainframe. Unlike the AIRS mainframe, AIRS Executive does not allow you to manipulate data or to custom-format reports, nor does it contain the extensive air pollution data found in AIRS.

The AIRS mainframe database contains the most comprehensive air pollution data in the world. The mainframe is housed at EPA's National Computer Center in Research Triangle Park, North Carolina. Any organization or individual with direct access to the EPA computer system can use AIRS to retrieve non-confidential air pollution data.

AIRS Executive Software and data files are available from the U.S EPA's gopher server, EPA's web server, EPA's anonymous ftp, and from EPA's electronic bulletin board system (TTNBBS).

3.5.4 AIRS Executive International
http://www.epa.gov/docs/airs/aexec.html

AIRS Executive International is a special version of AIRS Executive that contains information about ambient air pollution in nations that voluntarily provide data to the GEMS/AIR Programme sponsored by the United Nations World Health Organization. Nearly 50 nations are participants. The GEMS/AIR data is stored in the AIRS database, just like EPA data for the U.S. AIRS Executive International is an easy way to view summaries of GEMS/AIR pollution monitoring data. Emissions data is not included in the International Version at this time.

AIRS Executive International is updated quarterly. Updates are published on the 10th day of January, April, July, and October.

3.5.5 California's ARBIS
http://arbis.arb.ca.gov/program.htm

The ARBIS is designed to provide information on the Air Resources Board's (ARB) activities for industry, local air districts, ARB staff, and the general public. Figure 3.5.5 shows the ARB home page. In October 1995, the ARBIS contained the following features:

Business Assistance: General information on ARB's Business Assistance Program including several referral guides.

California Air Pollution Statutes and Regulations: Includes the District Rules Database, the California Air Pollution Control Laws (i.e., the "Blue Book"), and the Federal Clean Air Act for viewing, downloading, and/or keyword searching. Also, the Federal Clean Air Act area includes information on state and local activities related to the implementation of Titles III and V of the act.

Mobile Source Program: Includes information on regulations regarding zero emission vehicles, low emission vehicles, the on-board diagnostics program, and mobile source emission reduction credits.

Air Toxics Program: Includes the Toxic Updates, the AB2588 overview and fee regulation, all air toxic control measures, and the U.S. EPA Hazardous Air Pollutants Health Effects Notebook.

Disclaimer , Maintenance 1-12-96

 General Information

- CARB's Mission Statement
- The Board (Members, Agendas, Summaries, Transcripts)
- Press Releases
- CARB Divisions
- Local and Regional Air Pollution Control Agencies

 CARB Programs and Information Resources

- CARB Programs
- Public Information Office
- About ARBIS dial-up BBS
- WWW Usage statistics

 Links outside of CARB

Please send comments and suggestions to *webmaster@www.arb.ca.gov*

Figure 3.5.5

Consumer Products: Includes the consumer products regulations, ARB staff contact lists, and related regulatory activities for the Consumer Product Working Group and subgroups.

Reformulated Gasoline (RFG) Program: Includes all agendas and final summaries of the Advisory and the three Subcommittee activities plus a question and answer forum and information on how to get fact sheets and RFG Newsletter information.

CAPCOA Information/Air District Listings: Includes the listings of CAPCOA committees, air district contacts, and documents available from CAPCOA.

Compliance Program: Includes information regarding available program resources, the 1995 training schedule, and course descriptions and notices for future variance hearings.

Best Available Control Technology (BACT) Clearinghouse: The 1993 BACT Compilation document is available for download.

Transportation and Land Use Programs: The URBEMIS5 computer model and supporting files to operate the model are available for download.

Message Center: Allows users to post messages.

Board Activities: Includes board meeting agendas, summaries, and transcripts beginning January 1995.

News Releases: The ARBIS now makes available all ARB news releases.

In the future, the ARBIS is expected to include these additional features:

- Health Risk Assessment Software
- Equipment/Emissions Database
- Portable Equipment Program Database
- ARB Workshop Notices
- Local Air District Informational Area

The ARBIS may be accessed via modem by calling 916-322-2826. Make sure your communications parameters are set to 8-N-1. If you have a 9600 baud modem or greater, use the ANSI capabilities that are provided by the more recent modem software packages. Modems slower than 9600 baud will work with VT-100 or TTY terminal emulation.

If you have any questions regarding access to the ARBIS, contact the Business Assistance Hot Line at 1-800-ARB-HLP2, Raul Cisneros at 916-327-5978, or Bill Fell at 916-322-3260.

3.5.6 WRSIC
http://www.uwin.siu.edu/databases/wrsic/

This site contains the Selected Water Resources Abstracts database. The database is an international collection of water research compiled by WRSIC. Research selected to be abstracted includes research in the life, physical, and social sciences, as well as engineering and legal issues. This database covers the time period from 1967 to October 1993.

The database contains over 265,000 abstracts and citations. Each citation gives full bibliographic information so that you can then locate your titles of interest. Most can be found through normal library resources using the information provided in the citation.

Searching through this database is done through user-provided keywords. Given the tremendous size of this database, a carefully chosen set of keywords is needed to locate the types of abstracts you desire.

A list of the general air areas covered by the database is provided below:

- meteorology, geophysics, energy
- atmospheric technology
- fluid mechanics, physics
- climatology, mathematical modeling

3.5.7 NOAA's Data Set Catalog
http://www.esdim.noaa.gov/NOAA-Catalog

NOAA's Data Set Catalog is a forms-based tool that allows users to search for publicly available environmental data held by public and private sources throughout the world. However, the databases that you search when you search the Data Set Catalog do not hold the actual data; instead the databases document the existence, location, characteristics, and availability of environmental data. Where possible, the catalog links the data set description to the actual data, the data center where the data is located, or to an order form at the data center of interest.

3.5.8 The NCDC
http://www.ncdc.noaa.gov/

The NCDC provides online data access to its data sets, including data inventories, long-term climatological data sets, U.S. monthly precipitation data, monthly temperature data, and special sensor microwave/imager data sets.

Figure 3.5.8 shows the NCDC home page.

3.5.9 Environmental Guidance Documents
http://www.tis.eh.doe.gov/docs/egm/

The gopher site at the address given above provides guidance information regarding the Clean Air Act (CAA). Figure 3.5.9 in Appendix A contains a printout of the list of guidance documents available regarding the CAA.

3.5.10 Environmental Engineering Mailing List

ENVENG-L is a mailing list for the discussion of environmental engineering practice, education, and research. Environmental engineering is broadly defined to include waste treatment, water supply, air pollution control, solid waste management, noise and radiation protection, waste minimization, and related issues.

To subscribe, send an email message to *listproc@pan.cedar.univie.ac.at*. The message body should contain the following as the only text:
   subscribe enveng-l <your first name> <your last name>

The National Climatic Data Center (NCDC)

☐ What's New
☐ Products, Publications and Services
☐ On-Line Data Access
☐ Interactive Visualization of Climate Data
☐ Climate Research Programs
☐ World Data Center A for Meteorology
☐ National Oceanic and Atmospheric Administration (NOAA)

http://www.ncdc.noaa.gov/ncdc.html
Last updated 29 Nov 95 by webmaster@ncdc.noaa.gov
Please see the NCDC Contact Page if you have questions or comments.

Figure 3.5.8

3.5.11 Environmental Newsgroups

Newsgroups are a great place to find current discussions in a particular field and to find FAQs and pointers to web sites, mailing lists, and other resources. However, most of the newsgroups are unmoderated, leading to a lot of noise (irrelevant messages). Some of the newsgroups to check out include:

news:sci.environment   *news:sci.engr*
news:sci.engr.civil   *news:sci.engr.chem*
news:sci.chemistry   *news:sci.chem.analytical*
news:sci.engr.safety

3.6 HAZARDOUS WASTE INFORMATION

There are a large number of sites on the Internet that can provide useful hazardous waste information to environmental managers. This section describes databases that are available on the Internet or on BBSs.

California electronic information services are also described, for two reasons. First, the state is among the leaders in the computerized supply of information. And second, we wish to point out that useful information can be found at the state level. Obviously, this manual could not provide information regarding environmental electronic resources available in every state, but hopefully these examples provide you with some ideas about what kind of information your state might have online now or in the future.

The following databases are described in this section:

1. Resource Conservation and Recovery Information System (RCRIS)

2. RTK NET's Biennial Reporting System (BRS)

3. Comprehensive Environmental Response, Compensation and Liability Information System (CERCLIS)

4. RTK NET's Superfund National Priorities List (NPL)

5. EPA's Alternative Treatment Technology Information Center (ATTIC)

6. EPA's Clean-Up Bulletin Board (CLU-IN)

7. EPA's Vendor Information System for Innovative Treatment Technologies (VISITT)

8. EPA's VendorFacts BBS

9. The USGS's WRSIC

10. The ATSDR's Hazdat Database

11. EPA's OLS

12. Environmental Guidance Documents

13. California's DTSC

14. California's Integrated Waste Management Board (IWMB)

15. Environmental Engineering Mailing List

16. Environmental Newsgroups

17. The Department of Transportation's (DOT's) Research and Special Projects Administration's Hazardous Materials Information Exchange (HMIX) BBS

3.6.1 EPA's RCRIS

http://www.epa.gov/enviro/html/rcris/rcris_overview.html

Generators, transporters, treaters, storers, and disposers of hazardous waste are regulated by RCRA. RCRIS is primarily used to track handler permit or closure status, compliance with federal and state regulations, and cleanup activities. RCRIS contains data regarding handler names and addresses, hazardous waste categories and activities, owners and operators, and authorized waste handling methods.

3.6.2 RTK NET's BRS
telnet://198.3.148.6:23

RTK NET is used mostly by environmental and public interest groups. It is designed to provide easy public access to EPA's TRI information. RTK NET includes the BRS database which contains the RCRA biennial reporting information, including information about shipments and storage of hazardous waste.

3.6.3 EPA's CERCLIS
http://www.epa.gov/enviro/html/cerclis/cerclis_overview.html

CERCLIS is the official repository for site- and non-site-specific Superfund data. It contains information on hazardous waste site assessment and remediation from 1983 to the present. Each site in CERCLIS is either currently listed on the NPL or is being or has been investigated as a Superfund site.

3.6.4 RTK NET's NPL List
telnet://198.3.148.6:23

RTK NET is used mostly by environmental and public interest groups. It is designed to provide easy public access to EPA's TRI information. RTK NET includes the NPL Superfund list.

3.6.5 EPA's ATTIC

ATTIC is a comprehensive automated bibliographic reference that integrates existing hazardous waste data into a unified, searchable resource.

Telnet:	None at this time
Voice Number:	703-908-2137
Data Number:	703-908-2138
Modem Settings:	8 data bits, No parity, 1 stop bit, terminal emulation VT100 or VT/ANSI, full duplex

3.6.6 EPA's CLU-IN
telnet://clu-in.epa.gov

CLU-IN is a publicly accessible, online computer system that fosters technology transfer and facilitates communication among those involved in solid and hazardous waste cleanup. It can also be reached at:

Telnet:	*clu-in.epa.gov* or *134.67.99.13*
Voice Number:	301-589-8368
Data Number:	301-589-8366
Modem Settings:	8 data bits, No parity, 1 stop bit

3.6.7 EPA's VISITT

VISITT is a database developed by EPA to provide current information about innovative technologies designed to remediate groundwater or nonaqueous phase liquids (NAPL) in situ, and in soil, sludge, soil-matrix waste, natural sediments, and off-gas. VISITT is designed to contain detailed information to enable users to screen and assess remediation technologies quickly. VISITT does not include incineration, aboveground wastewater or groundwater treatment, solidification/ stabilization, or industrial waste treatment technologies.

VISITT enables the user to build queries according to the conditions at a site. Search criteria include: contaminant group, contaminant data, media, waste source, technology type, scale, vendor name, trade name, state/province, country, and business size.

VISITT files can be downloaded from a number of online sources. For instructions on downloading VISITT call 800-245-4505, or 703-883-8448.

Telnet:	*clu-in.epa.gov* or *134.67.99.13*
CLU-IN Data Number:	301-589-8366
Modem Settings:	8 data bits, No parity, 1 stop bit

VISITT can also by downloaded from AOL, the Defense Environmental Network for Information eXchange (DENIX), and EPA's ftp site: *ftp.epa.gov*. Other sites should also be available in the near future.

3.6.8 EPA's VendorFacts Bulletin

VendorFacts is designed to promote the use of innovative technologies for field analytical techniques and site characterization. It includes transportable technologies for on-site screening, characterizing, monitoring, and analysis of hazardous substances. Technologies used for collecting samples for off-site analysis or for monitoring industrial process waste streams are not included in the database. The system allows users to screen technologies by parameters such as contaminants, media, or development status.

VendorFacts files can be downloaded from CLU-IN. To do so, you need a password which can be obtained by filling out a VendorFacts registration questionnaire.

Telnet:	*clu-in.epa.gov* or *134.67.99.13*
Voice Number:	301-589-8368
Data Number:	301-589-8366
Modem Settings:	8 data bits, No parity, 1 stop bit

VendorFacts can also by downloaded from AOL, DENIX, and EPA's ftp site: *ftp.epa.gov*. Other sites should also be available in the near future.

3.6.9 WRSIC
http://www.uwin.siu.edu/databases/wrsic/

This site contains the Selected Water Resources Abstracts database. The database is an international collection of water research compiled by WRSIC. Research selected to be abstracted includes research in the life, physical, and social sciences, as well as engineering and legal issues. This database covers the time period from 1967 to October 1993.

The database contains over 265,000 abstracts and citations. Each citation gives full bibliographic information so that you can then locate your titles of interest. Most can be found through normal library resources using the information provided in the citation.

Searching through this database is done through user-provided keywords. Given the tremendous size of this database, a carefully chosen set of keywords is needed to locate the types of abstracts you desire.

A list of the general waste areas covered by the database is provided below:

- hazardous waste and effluent treatment
- pollution and waste management

3.6.10 Hazdat
http://atsdr1.atsdr.cdc.gov:8080/hazdat.html

Hazdat is a database developed and maintained by the ATSDR, developed to provide access to information on:

- the release of hazardous substances from Superfund sites
- the release of hazardous substances from emergency events
- the effects of hazardous substances on the health of human populations

The Superfund site information listed below for more than 1,300 sites is included in Hazdat. However, obtaining the information is a little more complicated than simply pointing and clicking. You will need to obtain certain software and set up your system appropriately before accessing this portion of the database. Instructions as to how to do so are provided by ATSDR.

- site characteristics
- site activities
- site events
- contaminants found
- contaminant media
- maximum concentration levels
- impact on population
- community health concerns
- ATSDR public health threat categorization
- ATSDR recommendations
- environmental fate of hazardous substances at the site/event
- exposure routes at the site/event
- physical hazards at the site/event

Internet Hazdat may also be queried directly from a web browser with Forms support by clicking on any one of the database queries listed at the Hazdat home page. Users can make site activity queries, site contaminant queries, toxicological profile queries, and public health statement text searches.

3.6.11 EPA's OLS

telnet://epaibm.rtpnc.epa.gov

EPA's OLS contains several databases including its Hazardous Waste database, which contains citations and summaries of key materials on hazardous waste. It can be reached at:

Telnet:	*epaibm.rtpnc.epa.gov*
Voice Number:	919-541-2777
Data Number:	919-549-0700 (2400)
	919-549-0720 (9600)
Modem Settings:	7 data bits, 1 stop bit, Even parity, half duplex

3.6.12 Environmental Guidance Documents

http://www.tis.eh.doe.gov/docs/egm

The gopher site at the address given above provides guidance information regarding the following waste issues:

- CERCLA (Figure 3.6.12A in Appendix A lists the documents available)
- Mixed Waste
- Radioactive Waste
- RCRA (Figure 3.6.12B in Appendix A lists the documents available)
- SARA (Figure 3.6.12C in Appendix A lists the documents available)

3.6.13 California's DTSC

http://www.cahwnet.gov/epa/dtsc.htm

California's DTSC is the agency that regulates hazardous materials in the state of California. Among the documents regarding hazardous waste that can be found at the DTSC web site are:

- Reorganization of Hazardous Waste Management Program
- List of Registered Hazardous Waste Transporters
- "Expired" Chemicals
- Non-RCRA Facilities/Offsite
- Petroleum Exclusion
- ID#s for Generators Using Milkruns
- Financial Assurance for Closure of PBR and CA Facilities

- Generator Inspection Checklist
- Checklist Guidance Document
- Contaminated Container Regulations
- Hazardous Waste Intro Info
- Hazardous Waste at Drycleaners
- Hazardous Waste & Printed Circuit Boards
- Hazardous Waste in Medical Offices
- Hazardous Waste in Paint Manufacturing
- Hazardous Waste in Pesticide Industry
- Hazardous Waste in Photoprocessing Industry

3.6.14 The IWMB Web Page

http://www.ciwmb.ca.gov/

The IWMB web page can be accessed through California EPA's web home page (the URL is shown above). The IWMB page provides its annual report, including:

- a list of all active and inactive landfills
- a list of used oil collection centers
- press releases

3.6.15 Environmental Engineering Mailing List

ENVENG-L is a mailing list for the discussion of environmental engineering practice, education, and research. Environmental engineering is broadly defined to include waste treatment, water supply, air pollution control, solid waste management, noise and radiation protection, waste minimization, and related issues.

To subscribe, send an email message to *listproc@pan.cedar.univie.ac.at*. The message body should contain the following as the only text:
   subscribe enveng-l <your first name> <your last name>

3.6.16 Environmental Newsgroups

Newsgroups are a great place to find current discussions in a particular field and to find FAQs and pointers to web sites, mailing lists, and other resources. However, most of the newsgroups are unmoderated, leading to a lot of noise (irrelevant messages). Some of the newsgroups to check out include:

news:sci.environment   *news:sci.engr*

news:sci.geo   *news:sci.geo.geology*

news:sci.geo.petroleum   *news:sci.geo.hydrology*

news:sci.engr.civil   *news:sci.engr.chem*

news:sci.chemistry   *news:sci.chem.analytical*

news:sci.engr.safety

3.6.17 DOT's HMIX Database
telnet://hmix.dis.anl.gov

The useful information on DOT's HMIX BBS is not particularly accessible to the public. However, it does provide very useful information to government employees. HMIX provides hazardous materials technical assistance and much more. It can also be reached at:

Voice Number: 800-752-6367
Data Number: 708-972-3275

3.7 RELEASE AND RISK INFORMATION

There are a large number of sites on the Internet that can provide useful release and risk information to environmental managers. This section describes the following databases that are available on the Internet or on BBSs:

1. RTK NET's Emergency Release Notification System (ERNS)

2. EPA's Toxic Release Inventory System (TRIS)

3. RTK NET's TRI data

4. EPA's Center for Exposure Assessment Modelling (CEAM)

5. The Fish and Wildlife Services's (FWS's) National Ecology Research Center (NERC)

6. Hazardous Materials Mailing List

7. Environmental Training Mailing List

8. Environmental Newsgroups

3.7.1 RTK NET's ERNS
telnet://198.3.148.6:23

RTK NET is used mostly by environmental and public interest groups. It is designed to provide easy public access to EPA's Toxic Release Inventory (TRI) information. RTK NET includes the ERNS database which provides details on reports of accidental releases as reported to the EPA. The accidental release information in this database covers the period from 1987 through 1992.

3.7.2 EPA's TRIS
http://www.epa.gov/enviro/html/tris/tris_overview.html

TRIS contains information about release and transfers of more than 300 toxic chemicals and compounds to the environment. TRIS stores release-transfer data hierarchically by facility, by year and chemical, and by medium of release (air, water, underground injection, land disposal, and off-site).

TRIS stores treatment and source-reduction data. TRIS also stores Standard Industrial Classification (SIC), EPA identification numbers (EPA ID), and pollution prevention data, e.g., recycling, energy recovery, treatment and disposal, and names and addresses of off-site transfer recipient facilities.

The number of chemicals that EPA requires reporting for has changed significantly over the years. Occasionally, currently listed chemicals may be delisted. Data for all chemicals is therefore not always available for every year.

3.7.3 RTK NET's TRI Data
telnet://198.3.148.6:23

RTK NET is used mostly by environmental and public interest groups. It is designed to provide easy public access to EPA's TRI information. It includes the TRI data for all years currently available beginning with 1987 (that is, 1987 to 1993).

3.7.4 EPA's CEAM
ftp://ftp.epa.gov/epa_ceam/wwwhtml/ceamhome.html

CEAM serves as the focal point for ORD's multimedia assessment modeling, ecological risk assessment, and distribution of related software products. For a current list of the CEAM software products distributed through the Internet, type in the URL above. The URL will take you to CEAM's home page. From there, click on the highlighted "software products" entry. From the CEAM software products page, you can click on a file in the left-hand column to download the corresponding DOS executable file. You can also click on the description on the right-hand column, and a page describing the corresponding software product will be displayed.

CEAM can also be reached at its BBS at:

Voice Number:	706-546-3549
	FTS-250-3549
Data Number:	706-546-3402
	FTS-250-3549

CEAM also provides a mailing list service. The mailing list is called CEAM-USERS. The mailing list broadcasts up-to-date information concerning CEAM software products, activities, and events. These include announcements for CEAM and non-CEAM supported workshops and training sessions; software product version releases, updates, and documentation information; hints on software installation, operation, application, problems, or enhancements; and the exchange of information quickly among users and between users and CEAM support personnel.

To subscribe, send an email message to the following administrative address: *listserver@unixmail.rtpnc.epa.gov*. In the subject line of the message, type **subscribe CEAM-USERS <Firstname><Lastname>**. No text is required in the body of the message. To unsubscribe, send a message to the administrative address, but type **unsubscribe CEAM-USERS** in the subject of the message. To broadcast a message, send your message to *CEAM-USERS@unixmail.rtpnc. epa.gov*. For help, type **help** in the subject of a message sent to the administrative address.

3.7.5 FWS's NERC

NERC maintains a Habitat Model Reference Library BBS which includes:

- habitat model information (species, author, geographic application)
- messages and discussions
- habitat models
- negotiation strategies
- habitat simulation
- temperature models
- computer utilities

The Habitat Model BBS-1 can be reached at:

Voice Number: 303-226-9293
Data Number: 303-226-9365

The NERC Habitat Model Reference Library can be reached at:

Voice Number: 303-226-9293
                 FTS-323-5293
Data Number: 303-226-9365
                 FTS-323-5365

3.7.6 Hazardous Materials Mailing List

This mailing list focuses on transportation, storage, and reporting of hazardous materials. To subscribe to this mailing list, send an email message with the subject blank to *listserv@cc.colorado.edu*. The message should read:
subscribe HAZMAT-L <firstname><lastname>

3.7.7 Environmental Training Mailing List

This unmoderated mailing list covers all aspects of environmental training, including needs assessments, selection of training topics, macro-designs, micro-designs, evaluation, followup, impact assessment, and project management. To subscribe to this mailing list, send an email message with the subject blank to *ListProc@Poniecki.Berkeley.edu*. The message should read:
Sub ETP-L <firstname><lastname>

Initial discussions will focus on Central and Eastern Europe, Russia, and the newly independent states.

3.7.8 Environmental Newsgroups

Newsgroups are a great place to find current discussions in a particular field and to find FAQs and pointers to web sites, mailing lists, and other resources. However, most of the newsgroups are unmoderated, leading to a lot of noise (irrelevant messages). Some of the newsgroups to check out include:

news:sci.environment   *news:sci.engr*
news:sci.geo   *news:sci.geo.geology*
news:sci.geo.petroleum   *news:sci.geo.hydrology*
news:sci.engr.civil   *news:sci.engr.chem*
news:sci.chemistry   *news:sci.chem.analytical*
news:sci.engr.safety

3.8 GENERAL ENVIRONMENTAL INFORMATION

There are a large number of sites on the Internet that can provide useful general environmental information to environmental managers. This section describes the following databases that are available on the Internet or on BBSs:

1. Environmental Agency Contacts

2. Online Library Catalogs

3. EPA's Envirofacts

4. EPA's Facilities Index System (FINDS)

5. RTK NET's Docket Database

6. RTK NET's 1990 Census Information

7. Environmental Guidance Documents Database

3.8.1 Environmental Agency Contacts

Agency home pages usually provide an organizational chart and/or a list of names, addresses, and phone numbers of contacts.

Other resources are also referenced on the web. For example, ACCESS EPA has over 300 entries of EPA and other private information sources such as hotlines, clearinghouses, etc. ACCESS EPA can be reached through EPA's OLS at:

Telnet:	*epaibm.rtpnc.epa.gov*
Voice Number:	919-541-2777
Data Number:	919-549-0700 (2400) or 919-549-0720 (9600)
Modem Settings:	7 data bits, 1 stop bit, Even parity, half duplex

3.8.2 Online Library Catalogs

Most academic institutions and many large organizations now catalog their library holdings electronically. Many of these library catalogs are available online (though not necessarily through the Internet). Three examples of online library catalogs are provided below. Phoning the institution's library will generally allow you to determine if its holdings are accessible to the public and if it has an online catalog of its holdings.

GLADIS — U.C. Berkeley's Online Catalog
telnet://gopac.berkeley.edu   or   *telnet://gladis.berkeley.edu*

All books, maps, and manuscripts acquired since 1977 and all periodicals owned by the Main, Moffitt, and Bancroft Libraries are cataloged and available using GLADIS. Older materials are also being added. Keyword and boolean searches are not available.

After connecting to the telnet address, you will be prompted to type "EXIT" or "GOPAC." Type "GOPAC" and you will be connected to the GLADIS catalog.

SOCRATES — Stanford's Online Catalog
telnet://forsythetn.stanford.edu

SOCRATES catalogs all books and journals in the Stanford collection. After connecting to the telnet address, type **SOCRATES** at the prompt and you will be connected to the SOCRATES catalog.

EPA's OLS
telnet://epaibm.rtpnc.epa.gov

OLS's National Catalog provides citations and summaries of environmentally related topics encompassing biology, chemistry, ecology, and other basic sciences and EPA reports distributed through the National Technical Information Service (NTIS).

3.8.3 Envirofacts
http://www.epa.gov/enviro/html/ef_home.html

Envirofacts is a compilation of EPA databases available via the web. Figure 3.8.3A shows the Envirofacts home page. Envirofacts combines data extracted from four major EPA databases into a single relational database. The databases that comprise Envirofacts are:

- PCS   (See sections 3.3.1 and 3.3.2)
- RCRIS   (See section 3.6.1)
- CERCLIS   (See section 3.6.3)
- TRIS   (See sections 3.7.1 and 3.7.2)

Envirofacts also includes:

- FINDS   (See section 3.8.4)
- EMCI   (See section 3.2.6)

These two databases link data from the four systems by providing facility and chemical identification numbers common to each system.

Envirofacts is updated monthly and contains data that is available under the Freedom of Information Act (FOIA). No enforcement or budget-sensitive data is contained in the database.

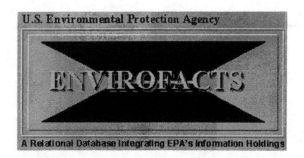

Welcome to the **Information Guide** for ENVIROFACTS.

ENVIROFACTS is a relational database that integrates data from four major EPA program systems: RCRIS, TRIS, CERCLIS, and PCS. It contains data, updated monthly, that is available under the Freedom of Information Act (FOIA). No enforcement or budget sensitive information is contained in this database.

Responses to frequently asked questions or comments are posted in the <u>Frequently Asked Questions (FAQ)</u> area. Improvements, changes and updates are posted in the <u>What's New</u> section. Users should refer to these sections first if there are any questions about changes.

The databases included in ENVIROFACTS are:

Permit Compliance System (PCS): Data on more than 75,000 water-discharge permits including permit issuance, permit limits, monitoring data and other data pertinent to facilities with permits.

Comprehensive Environmental Response, Compensation and Liability Information System (CERCLIS): Superfund data on hazardous waste site assessment and remediation, including data on active sites from point of discovery to listing on the National Priorities List through completion of remedial and response actions.

Toxic Release Inventory System (TRIS): Data about the release and transfer of more than 300 toxic chemicals and compounds by medium of release (air, water, underground injection, land disposal and off-site), reported by over 33,000 submitters

Resource Conservation and Recovery Information System (RCRIS): Data used to track handler permit or closure status, compliance with Federal and State regulations, and cleanup activities, including hazardous waste generation and management data on over 450,000 facilities and transporters.

In addition, the following databases are included to provide integration within ENVIROFACTS:

Facility Index System (FINDS): A central inventory of over 675,000 facilities regulated/monitored by program offices within the EPA.

Figure 3.8.3A

A packaged query form allows retrieval of facility information by city and state or zip code. It only provides a list of forms and their addresses; it does not provide information about chemicals and releases.

To get more detailed information about chemicals and releases at a site, you have to connect directly to Envirofacts and structure your own queries. The FAQ file describes how this can be done in more detail. Figure 3.8.3B shows the relevant FAQ page.

Connecting directly to Envirofacts requires *SQL*NET*, a program that uses a TCP/IP connection to connect your database program (such as Paradox, 123, or QuattroPro) to the EPA's database. Your database program may also require Open Database Connectivity drivers for an Oracle Version 7 database. The *SQL*NET* software is available from Oracle (800-633-0586) for $150.

The database is currently operating in pilot mode and EPA welcomes comments and suggestions. Enhancements to the forms are being designed to provide additional flexibility.

3.8.4 Facilities Index System (FINDS)
http://www.epa.gov/enviro/html/finds

FINDS is a central inventory of facilities regulated or monitored by the different programs within the EPA. The system functions as a repository for facilities monitored by EPA, a repository of spatial data (i.e., latitude and longitude data) for those facilities, and as an integrator for facilities monitored by more than one program office. The information in FINDS includes:

- facility name, street address, city, state, county, and ZIP code to define the location of a facility
- business information such as Standard Industrial Classification (SIC) codes, DUNS Numbers, ownership of facilities, etc.
- spatial data for the facilities

FINDS also provides a cross reference between the FINDS Facility ID and System Identification Codes used to identify facilities in program office systems (e.g., Handler ID in RCRIS, NPDES Number in PCS, EPA ID in CERCLIS, and TRI Facility ID in TRIS).

 Frequently Asked Questions

Question: Is the ENVIROFACTS data available for downloading via anonymous ftp?

Answer: The ENVIROFACTS database is only available for on-line querying.

Question: How can I access the ENVIROFACTS database via the internet and what software do I need to do so?

Answer: The ENVIROFACTS database can be accessed by any application that can communicate with an ORACLE database. There are several database querying and reporting products available. In the Microsoft Windows environment, products are available from a variety of sources, for example Borland, Gupta, IQ Software, Microsoft, ORACLE Corp., Q+E, etc. These products represent a range of capabilities from point and click querying to direct entry of Structured Query Language (SQL) statements.

In addition to a database querying or reporting application, you will also need:

1. SQL*NET for TCP/IP from ORACLE corporation; and
2. TCP/IP access to the internet.

[Microsoft Windows based applications from vendors other than ORACLE Corporation, may also require Open Database Connectivity (ODBC) drivers for an ORACLE Version 7 database from the vendor of the application.]

A block diagram of this arrangement is shown below:

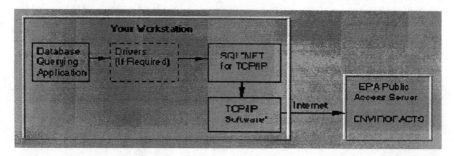

*The TCP/IP software may be 'loaded' for you by your Internet access provider when you initiate your connection. This happens, for example, if you have a SLIP connection to the Internet.

Once the necessary software has been installed on your workstation, you will need to provide the connection and identification information about ENVIROFACTS to your application. While the exact format of this information will vary from product to product, the following should suffice for all applications:

Figure 3.8.3B

Database server: earth1.epa.gov

Database name: enviro

Username: enviro

Password: enviro

Question: The forms only provide a list of facilities and their addresses. I thought they would give me information about chemicals and releases. How can I get this information?

Answer: The forms based queries are initially intended to provide a list of facilities. You can obtain detailed information about these facilities by connecting to ENVIROFACTS and executing your own queries using a database.

However, enhancements to these forms are being designed to provide additional flexibility. Many good suggestions have been submitted via the Feedback forms and these will be taken into consideration in designing the enhancements.

Question: Can I telnet to earth1.epa.gov and connect to ENVIROFACTS?

Answer: The earth1 server is not available for direct log-in by users.

 **ENVIROFACTS**

◩ **Metadata**

    Identification Information

    Data Quality Information

    Distribution Information

    Metadata Reference Information

◩ **Feedback Form**

◩ **Home Page**

Figure 3.8.3B

RTK NET's FINDS
telnet:///198.3.148.6:23

RTK NET is used mostly by environmental and public interest groups. It is designed to provide easy public access to EPA's Toxic Release Inventory (TRI) information. RTK NET includes EPA's FINDS database to assist in the identification of industrial activities under other than TRI reporting requirements.

3.8.5 RTK NET's Docket Database
telnet:///198.3.148.6:23

RTK NET includes the Docket Database which provides information about all civil cases brought by the EPA.

3.8.6 RTK NET's Census Information
telnet:///198.3.148.6:23

RTK NET includes some select 1990 Census information.

3.8.7 Environmental Guidance Documents
http://www.tis.eh.doe.gov/docs/egm/

The gopher site at the address given above provides guidance information regarding the following issues:

- Department of Energy (DOE) comments to other agencies
- Environmental Surveys
- Federal Facilities
- General Information
- Training
- the National Environmental Policy Act (NEPA)
- the Toxic Substances Control Act (TSCA) (Figure 3.8.7 in Appendix A lists the TSCA guidance documents available)

4

Environmental Information on the Internet for a Fee

Many commercial information services are also available via the Internet. These services are generally billed to a credit card. You may set up the account via the Internet, telephone, or fax.

If you are concerned about security over the Internet (and you should be, if you are typing in your credit card number), then make sure you are using the latest version of Netscape and Mosaic or similar programs that offer secure encrypted transmissions of data. In Netscape, a graphic key at the bottom left of the screen indicates whether or not you are connected to a compatible secure web page.

Although there is a wealth of free information that you can obtain via the Internet, it is usually from the government. Most newspapers, journals, and other publications are not freely available or do not provide the ability to search through current or past articles.

The following list of commercial information services is not comprehensive, but each provides potentially useful environmental information.

4.1 COUNTERPOINT

Counterpoint: *http://www.counterpoint.com*

Counterpoint is a division of Thompson Legal Publishing Company and provides access to the following databases:

- U.S. Federal Register and Archives - Updated Daily
- U.S. Commerce Business Daily - Updated Daily

- U.S. Code of Federal Regulations - Updated Monthly
- State Environmental Regulations - Updated Monthly
- Nuclear Information Service - DOE Orders and NRC Information
- TA5RUS Federal Government Contracting Information
- Americans with Disabilities Act Information

4.2 DUN & BRADSTREET

Dun & Bradstreet Information Services: *http://www.dbisna.com*

Dun & Bradstreet offers a Business Background Report which provides information on a company's history, business, background of its management, special events and recent newsworthy items Dun & Bradstreet has learned of, and a business operations overview. A report can be ordered using a credit card through the web for $20. Once your order is processed and the credit card authorized, your report is sent to you.

Dun & Bradstreet's database contains information on 11 million companies and is searchable by company name and state or telephone number.

4.3 KNIGHT-RIDDER/DIALOG

Knight-Ridder: *http://www.dialog.com*

Knight-Ridder is the owner of the Dialog and Datastar online information services.

Dialog provides access to current and past issues of hundreds of magazines and journals, most of them full text, through:

- BNA Daily News from Washington
- Chemical Engineering and Biotechnology Abstracts
- Energyline
- Energy Science & Technology
- Enviroline
- Environmental Bibliography

- Federal Register
- Federal Research in Progress
- Geoarchive
- Geobase
- Medline
- National Technical Information Service
- Pollution Abstracts
- Registry of Toxic Effects of Chemical Substances
- Toxline
- TRIS

Datastar provides access to more than 400 databases from a wide variety of subject areas including specialized pharmaceutical, biomedical, and health care information. The coverage, combined with the Datastar searching capabilities, make it a powerful online system.

Information retrieved from Datastar ranges from listings of citations with bibliographies to full-text journal or periodical articles.

A small sampling of the sources available from Datastar includes:

- MEDITEC: Biomedical Engineering
- Chemical Business NewsBase
- Chemical Engineering & Biotechnology Abstracts
- CA Search: Chemical Abstracts
- Chem Sources Company Directory
- Dissertation Abstracts Online
- ENVIROLINE
- FT Reports: Energy & Environment
- Hazardous Substances Databank
- HSELINE: Health & Safety
- KIRK Other Encyclopedia of Chemical Technology
- MEDLINE
- NTIS: National Technical Information Service
- Pollution Abstracts
- RTECS: Toxic Effects of Chemicals
- SciSearch: Science Citation Index
- Toxline

4.4 LEXIS-NEXIS

Lexis-Nexis: *http://www.lexis-nexis.com*

Lexis-Nexis provides access to a huge database of newspapers, laws, regulations, journals, and other information sources. Lexis-Nexis specializes in providing information to lawyers. Most of its business products, however, are tailored towards financial information and entrepreneurs.

Quick-Check, a Windows-based program, is designed to easily retrieve information from Lexis-Nexis in four categories:

News: A database of more than 2,300 news sources, including 100 local newspapers, major international newspapers, and magazines.

Company and Industry Information: Access to thousands of brokerage reports, company profiles, and reports from more than 140 investment banks.

Financial Reports: Provides access to annual reports, financial reports, 10-Ks, and 10-Qs by company name or ticker symbol.

Geographical: Provides economic and current political information on hundreds of countries.

Lexis-Nexis Small Business Service provides articles about operating your own business from more than 2,400 databases containing full-text news and business information.

4.5 WEST PUBLISHING

West Publishing: *http://www.westpub.com*

West Publishing is another information behemoth, and, like Lexis-Nexis, its specialty is providing information to lawyers. However, its access to full-text databases of news and journals can be very useful for the environmental professional.

Westlaw, one of West Publishing's services, offers access to the public records databases of Information America and TRW REDI Property Data, which help to identify information about people, businesses, and property.

Westlaw also offers access to Dialog, Dun & Bradstreet (discussed above) and Dow Jones news services, including the Wall Street Journal and Barron's. More than 5,000 databases are available, including state and federal statutes, federal regulations, journals, and periodicals.

And, should you need an attorney, West Publishing has a free directory of lawyers available at its web site.

4.6 EPA'S SOLID WASTE MANAGEMENT BBS

The Solid Waste Management BBS is a fee-based system. It can be reached at:

Voice Number:	800-677-9424
Data Number:	301-585-0204

4.7 EPA'S NPIRS

The National Pesticide Information Retrieval System (NPIRS) is a fee service available through the EPA. To use the databases, you need to set up a subscription to obtain a user identification and password. In October 1995, basic subscription rates were $300 a year.

NPIRS is a set of six databases which provide information to subscribers on pesticide products registered by EPA. Registration support documents, commodity/tolerance data, materials handling documents, and state product registration data are provided.

Voice Number:	317-494-6614
Data Number:	317-463-3262

Glossary

ASCII American Standard Code for Information Interchange. This is the standard for the 128 codes used to represent numbers and letters so that different computers can exchange information.

ATM Asynchronous Transfer Mode.

Bandwidth The capacity of any given data connection, whether it be a modem, phone line, or digital line. This capacity is generally measured in bits per second. Plain text uses much less bandwidth than graphics, sound, or video.

Baud Commonly used to mean bits per second, but actually refers to how many times the modem carrier signal shifts values.

BBS Bulletin Board System. This is generally a stand-alone computer system that can usually be reached only by making a telephone call.

BinHex Binary Hexadecimal. This is a format for converting non-ASCII files into ASCII so that they can be sent via Internet email, which generally can handle only ASCII. See also *uuencode*.

Bit Binary Digit. The smallest unit of data that a computer can understand. Represented by a one or zero. It takes eight bits (also known as a byte) to represent a single ASCII character.

Byte A set of bits (usually eight) that represents a single ASCII character.

Emoticon A set of standard characters commonly used to express emotions in messages. For example :) and :([look sideways to see the faces].

FAQ   Frequently Asked Questions. A FAQ is a document that tries to answer the most frequently asked questions about a specific subject. FAQs are commonly posted in relevant newsgroups and are stored at *ftp://rtfm.mit.edu.*

GIF   Graphic Interchange Format. A common format used to represent graphics. GIF graphics are hampered by their ability to display limited colors. Photos in GIF format generally take up much more space than photos in JPG format.

HTML   Hypertext Markup Language. The language used for web documents. HTML is represented by ASCII characters using special tags to define font size and type, graphic files, sound files, etc.

http   Hypertext Transfer Protocol. The protocol used by browsers and web servers to communicate with each other. Using http, text, sound, graphics, and video files can be transferred from one place to another.

ISDN   Integrated Services Digital Network. ISDN is a digital protocol for moving information over standard telephone lines. Widely available in California and quickly becoming available throughout the U.S., it provides two 64,000 bit per second channels and one 16,000 bit per second channel. The two 64,000 bit per second channels can be combined for a 128,000 bit per second throughput.

ISP   Internet Service Provider. An Internet service provider is an organization or company that provides a connection to the Internet either directly or via a telephone line connection.

JPG or JPEG   A standardized compression format for storing images. JPEG stands for Joint Photographic Experts Group, the name of the committee that wrote the standard. JPEG can store images such as photos, naturalistic artwork, or similar materials in files much smaller than GIFs. It can display many more colors than GIFs, but it is not good for line drawings, letterings, and simple cartoons.

K or Kb   See *kilobyte*.

Kilobyte   Commonly used to mean 1,000 bytes, but technically represents 1,024 bytes.

Link   A bit of text or a graphic that you point to (or highlight) and click on to move from page to page on the web.

MIME   Multipurpose Internet Mail Extensions. A format for attaching files to email. Not widely used by commercial online services such as CompuServe or AOL.

Modem   Modulator-Demodulator. A device that connects your computer to other computers via a telephone line.

Newbie   Internet slang for a new user. Often derided by veteran Internet users, newbies give themselves away by posting inappropriate messages or questions that could have been answered by reading a FAQ.

Page   The basic unit of information on the web. May contain text, graphics, sound, video, or some combination thereof. A home page is usually the starting point for a service or organization.

PDF   Portable Document Format. A type of file created by Adobe Acrobat. The Adobe Acrobat reader can be downloaded for free and used to view or print documents exactly as they were created, including tables, graphics, fonts, etc.

PPP   Point to Point Protocol. A protocol that allows a computer with a modem to use normal telephone lines to make a TCP/IP connection and become part of the Internet. A PPP or SLIP connection is generally required for home computer users to use browsers such as Netscape or Mosaic.

SLIP   Serial Line Internet Protocol. A protocol that allows a computer with a modem to use normal telephone lines to make a TCP/IP connection and become part of the Internet. A PPP or SLIP connection is generally required for home computer users to use browsers such as Netscape or Mosaic. SLIP is an older protocol and is being replaced by PPP.

TCP/IP Transmission Control Protocol/Internet Protocol. These are the protocols of the Internet. They are available to every type of computer system and are required for systems that are connected as part of the Internet. Also required to use Netscape or Mosaic. TCP Manager and Trumpet Winsock provide TCP/IP for PC users and MacTCP provides TCP/IP for Macintosh users.

Thread A series of messages concerning the same subject, formed by people responding to other messages. The responding messages will usually be denoted by *RE:* in the subject line of the message.

Uudecode Unix to Unix decode. This is a program for converting ASCII files (that have been converted by uuencode) into non-ASCII so that they can be used by the recipient.

Uuencode Unix to Unix encode. This is a program for converting non-ASCII files into ASCII so that they can be sent via Internet email, which generally can handle only ASCII. See also *BinHex*.

Veronica Very Easy Rodent-Oriented Net-wide Index to Computerized Archives. A tool for searching through gopherspace.

WWW or web World Wide Web. This is the colorful and lively part of the Internet characterized by its recent high rate of growth. The web has the ability to display text, graphics, and video. Sound can also be included in web pages.

Appendix A

Common Name: Acetone
CAS Number: 67-64-1
DOT Number: UN 1090
Date: January, 1986

— — — — — — — — — — — — —

HAZARD SUMMARY
* Acetone can affect you when breathed in and by passing through your skin.
* Exposure to high concentrations can cause you to become dizzy, lightheaded, and to pass out.
* Contact can irritate the skin. Repeated exposure may cause dryness.
* Contact can cause severe skin burns.
* Exposure can irritate the eyes, nose, and throat.
* Acetone is a FLAMMABLE LIQUID and a FIRE HAZARD.

IDENTIFICATION
Acetone is a colorless liquid with a sweet odor. It is used as a solvent and to manufacture other chemicals.

REASON FOR CITATION
* Acetone is on the Hazardous Substance List because it is regulated by OSHA and cited by ACGIH, NIOSH, NFPA, DOT and EPA.
* This chemical is on the Special Health Hazard Substance List because it is FLAMMABLE.
* Definitions are attached.

HOW TO DETERMINE IF YOU ARE BEING EXPOSED
* Exposure to hazardous substances should be routinely evaluated. This may include collecting air samples. Under OSHA 1910.20, you have a legal right to obtain copies of sampling results from your employer. If you think you are experiencing any work related health problems, see a doctor trained to recognize occupational diseases. Take this Fact Sheet with you.
* ODOR THRESHOLD = 13 ppm.
* The odor threshold only serves as a warning of exposure. Not smelling it does not mean you are not being exposed.

WORKPLACE EXPOSURE LIMITS
OSHA: The legal airborne permissible exposure limit (PEL) is 1,000 ppm averaged over an 8 hour work shift. NIOSH: The recommended airborne exposure limit is 250 ppm averaged over a 10 hour workshift.

ACGIH: The recommended airborne exposure limit is 750 ppm averaged over an 8 hour workshift and 1,000 ppm as a STEL (short term exposure limit).

* The above exposure limits are for air levels only. When skin contact also occurs, you may be overexposed, even though air levels are less than the limits listed above.

WAYS OF REDUCING EXPOSURE
* Where possible, enclose operations and use local exhaust ventilation at the site of chemical release. If local exhaust ventilation or enclosure is not used, respirators should be worn.
* Wear protective work clothing.
* Wash thoroughly immediately after exposure to Acetone and at the end of the workshift.
* Post hazard and warning information in the work area. In addition, as part of an ongoing education and training effort communicate all information on the health and safety hazards of Acetone to potentially exposed workers.

Figure 3.2.1 EPA's Chemical Substance Database:
Acetone (1 of 7 pages)

This Fact Sheet is a summary source of information of all potential and most severe health hazards that may result from exposure. Duration of exposure, concentration of the substance and other factors will affect your susceptibility to any of the potential effects described below.

— — — — — — — — — — — — — — — — —

HEALTH HAZARD INFORMATION

Acute Health Effects
The following acute (short term) health effects may occur immediately or shortly after exposure to Acetone:

* Contact can irritate the eyes or skin.
* Exposure can irritate the eyes, nose, and throat.
* High concentrations can cause you to become dizzy, lightheaded, and to pass out.

Chronic Health Effects
The following chronic (long term) health effects can occur at some time after exposure to Acetone and can last for months or years:

Cancer Hazard
* According to the information presently available to the New Jersey Department of Health, Acetone has not been tested for its ability to cause cancer in animals.

Reproductive Hazard
* According to the information presently available to the New Jersey Department of Health, Acetone has not been tested for its ability to adversely affect reproduction.

Other Long Term Effects
* High exposure may damage the liver and kidneys.
* Repeated skin contact with the liquid can cause dryness and irritation of the skin.
* Long term exposure can cause chronic nose and throat irritation.
* This chemical has not been adequately evaluated to determine whether brain or other nerve damage could occur with repeated exposure. However, many solvents and other petroleum based chemicals have been shown to cause such damage. Effects may include reduced memory and concentration, personality changes (withdrawal, irritability), fatigue, sleep disturbances, reduced coordination, and/or effects on nerves supplying internal organs (autonomic nerves) and/or nerves to the arms and legs (weakness, "pins and needles").

MEDICAL TESTING

If symptoms develop or overexposure is suspected, the following may be useful:

* Acetone can be measured in the blood, urine, and expired air, and has been used as an index of exposure.
* Interview for brain effects, including recent memory, mood (irritability, withdrawal), concentration, headaches, malaise and altered sleep patterns. Consider cerebellar, autonomic and peripheral nervous system evaluation. Positive and borderline individuals should be referred for neuropsychological testing.

Any evaluation should include a careful history of past and present symptoms with an exam. Medical tests that look for damage already done are not a substitute for controlling exposure.

Request copies of your medical testing. You have a legal right to this information under OSHA 1910.20.

Figure 3.2.1 EPA's Chemical Substance Database:
Acetone (2 of 7 pages)

WORKPLACE CONTROLS AND PRACTICES

Unless a less toxic chemical can be substituted for a hazardous substance, ENGINEERING CONTROLS are the most effective way of reducing exposure. The best protection is to enclose operations and/or provide local exhaust ventilation at the site of chemical release. Isolating operations can also reduce exposure. Using respirators or protective equipment is less effective than the controls mentioned above, but is sometimes necessary.

In evaluating the controls present in your workplace, consider: (1) how hazardous the substance is, (2) how much of the substance is released into the workplace and (3) whether harmful skin or eye contact could occur. Special controls should be in place for highly toxic chemicals or when significant skin, eye, or breathing exposures are possible.

In addition, the following controls are recommended:

* Where possible, automatically pump liquid Acetone from drums or other storage containers to process containers.
* Specific engineering controls are recommended for this chemical by NIOSH. Refer to the NIOSH criteria document: Ketones #78 173.

Good WORK PRACTICES can help to reduce hazardous exposures. The following work practices are recommended:
* Workers whose clothing has been contaminated by Acetone should change into clean clothing promptly.
* Contaminated work clothes should be laundered by individuals who have been informed of the hazards of exposure to Acetone.
* On skin contact with Acetone, immediately wash or shower to remove the chemical. At the end of the workshift, wash any areas of the body that may have contacted Acetone, whether or not known skin contact has occurred.
* Do not eat, smoke, or drink where Acetone is handled, processed, or stored, since the chemical can be swallowed. Wash hands carefully before eating or smoking.

PERSONAL PROTECTIVE EQUIPMENT

WORKPLACE CONTROLS ARE BETTER THAN PERSONAL PROTECTIVE EQUIPMENT. However, for some jobs (such as outside work, confined space entry, jobs done only once in a while, or jobs done while workplace controls are being installed), personal protective equipment may be appropriate.

The following recommendations are only guidelines and may not apply to every situation.

Clothing
* Avoid skin contact with Acetone. Wear solvent resistant gloves and clothing. Safety equipment suppliers/manufacturers can provide recommendations on the most protective glove/clothing material for your operation.
* All protective clothing (suits, gloves, footwear, headgear) should be clean, available each day, and put on before work.
* ACGIH recommends Butyl Rubber as a protective material.

Eye Protection
* Wear splash proof chemical goggles and face shield when working with liquid Acetone, unless full facepiece respiratory protection is worn.

Figure 3.2.1 EPA's Chemical Substance Database:
Acetone (3 of 7 pages)

Respiratory Protection
IMPROPER USE OF RESPIRATORS IS DANGEROUS. Such equipment should only be used if the employer has a written program that takes into account workplace conditions, requirements for worker training, respirator fit testing and medical exams, as described in OSHA 1910.134.

* Where the potential exists for exposures near or over 250 ppm, use a MSHA/NIOSH approved full facepiece respirator with an organic vapor cartridge/canister. Increased protection is obtained from full facepiece powered air purifying respirators.
* If while wearing a filter, cartridge or canister respirator, you can smell, taste, or otherwise detect Acetone, or in the case of a full facepiece respirator you experience eye irritation, leave the area immediately. Check to make sure the respirator to face seal is still good. If it is, replace the filter, cartridge, or canister. If the seal is no longer good, you may need a new respirator.
* Be sure to consider all potential exposures in your workplace. You may need a combination of filters, prefilters, cartridges, or canisters to protect against different forms of a chemical (such as vapor and mist) or against a mixture of chemicals.
* Where the potential for higher exposures exists, use a MSHA/NIOSH approved supplied air respirator with a full facepiece operated in the positive pressure mode or with a full facepiece, hood, or helmet in the continuous flow mode, or use a MSHA/NIOSH approved self contained breathing apparatus with a full facepiece operated in pressure demand or other positive pressure mode.
* Exposure to 20,000 ppm is immediately dangerous to life and health. If the possibility of exposures above 20,000 ppm exists, use a MSHA/NIOSH approved self contained breathing apparatus with a full facepiece operated in continuous flow or other positive pressure mode.

HANDLING AND STORAGE

* Prior to working with Acetone you should be trained on its proper handling and storage.
* Acetone must be stored to avoid contact with OXIDIZING MATERIALS (such as PEROXIDES, CHLORATES, PERCHLORATES, NITRATES, and PERMANGANATES) and ACIDS, since violent reactions occur.
* Store in tightly closed containers in a cool, well ventilated area away from HEAT, SPARKS, and FLAME.
* Sources of ignition, such as smoking and open flames, are prohibited where Acetone is used, handled, or stored in a manner that could create a potential fire or explosion hazard.
* Metal containers involving the transfer of 5 gallons or more of Acetone should be grounded and bonded. Drums must be equipped with self closing valves, pressure vacuum bungs, and flame arresters.
* Use only non sparking tools and equipment, especially when opening and closing containers of Acetone.

Figure 3.2.1 EPA's Chemical Substance Database:
Acetone (4 of 7 pages)

Common Name: Acetone
DOT Number: UN 1090
DOT Emergency Guide code: 26
CAS Number: 67-64-1

— — — — — — — — — — — — — — —

Hazard rating	NJ DOH	NFPA
FLAMMABILITY	-	3
REACTIVITY	-	0

— — — — — — — — — — — — — — —

POISONOUS GASES ARE PRODUCED IN FIRE
CONTAINERS MAY EXPLODE IN FIRE

— — — — — — — — — — — — — — —

Hazard Rating Key: 0=minimal; 1=slight; 2=moderate; 3=serious; 4=severe

FIRE HAZARDS
* Acetone is a FLAMMABLE LIQUID.
* Use dry chemical, CO2, or alcohol foam extinguishers. Use water to keep fire exposed containers cool.
* POISONOUS GASES ARE PRODUCED IN FIRE,.
* CONTAINERS MAY EXPLODE IN FIRE.
* If employees are expected to fight fires, they must be trained and equipped as stated in OSHA 1910.156.

SPILLS AND EMERGENCIES
If Acetone is spilled or leaked, take the following steps:

* Restrict persons not wearing protective equipment from area of spill or leak until cleanup is complete.
* Remove all ignition sources.
* Ventilate area of spill or leak.
* Absorb liquids in vermiculite, dry sand, earth, or a similar material and deposit in sealed containers.
* Keep Acetone out of a confined space, such as a sewer, because of the possibility of an explosion, unless the sewer is designed to prevent the build up of explosive concentrations.
* It may be necessary to contain and dispose of Acetone as a HAZARDOUS WASTE. Contact your state Environmental Program for specific recommendations.

— — — — — — — — — — — — — — — — — — — —

FOR LARGE SPILLS AND FIRES immediately call your fire department.

— — — — — — — — — — — — — — — — — —

FIRST AID

POISON INFORMATION

Eye Contact
* Immediately flush with large amounts of water for at least 15 minutes, occasionally lifting upper and lower lids.

Figure 3.2.1 EPA's Chemical Substance Database:
Acetone (5 of 7 pages)

Skin Contact
* Quickly remove contaminated clothing. Immediately wash contaminated skin with large amounts of soap and water.

Breathing
* Remove the person from exposure.
* Begin rescue breathing if breathing has stopped and CPR if heart action has stopped.
* Transfer promptly to a medical facility.

PHYSICAL DATA

Vapor Pressure: 266 mm Hg at 77oF (25oC)
Flash Point: 1.4oF (17.0oC)
Water Solubility: Miscible

OTHER COMMONLY USED NAMES
Chemical Name:
2-Propanone

Other Names and Formulations:
Dimethylformaldehyde; Dimethyl Ketone; Methyl Ketone.
— — — — — — — — — — — — — — — — — —

Not intended to be copied and sold for commercial purposes.
— — — — — — — — — — — — — — — — — —

NEW JERSEY DEPARTMENT OF HEALTH
Right to Know Program CN 368, Trenton, NJ 08625-0368

— — — — — — — — — — — — — — — — — —
— — — — — — — — — — — — — — — — — —

ECOLOGICAL INFORMATION

Acetone is a colorless, flammable liquid with a somewhat aromatic odor. It is widely used as a solvent for paint, varnish, lacquers, inks, adhesives, and heatseal coatings. It is also used extensively as a chemical intermediate in the production of pharmaceuticals and plastic or resin materials. It may enter the environment from industrial or municipal waste treatment plant discharges or spills.

ACUTE (SHORT-TERM) ECOLOGICAL EFFECTS

Acute toxic effects may include the death of animals, birds, or fish, and death or low growth rate in plants. Acute effects are seen two to four days after animals or plants come in contact with a toxic chemical substance.

Acetone has slight acute toxicity to aquatic life. Acetone has caused membrane damage, size decrease, and germination decrease in various agricultural and ornamental crops. Insufficient data are available to evaluate or predict the short-term effects of acetone to birds and land animals.

CHRONIC (LONG-TERM) ECOLOGICAL EFFECTS

Chronic toxic effects may include shortened lifespan, reproductive problems, lower fertility, and changes in appearance or behavior. Chronic effects can be seen long after first exposure(s) to a toxic chemical.

Figure 3.2.1 EPA's Chemical Substance Database:
Acetone (6 of 7 pages)

Acetone has slight chronic toxicity to aquatic life. Insufficient data are available to evaluate or predict the long-term effects of acetone to plants, birds, or land animals.

WATER SOLUBILITY

Acetone is highly soluble in water. Concentrations of 1,000 milligrams and more will mix with a liter of water.

DISTRIBUTION AND PERSISTENCE IN THE ENVIRONMENT

Acetone is slightly persistent in water, with a half-life of between 2 to 20 days. The half-life of a pollutant is the amount of time it takes for one-half of the chemical to be degraded. About 50% of acetone will eventually end up in air; the rest will end up in the water.

BIOACCUMULATION IN AQUATIC ORGANISMS

Some substances increase in concentration, or bioaccumulate, in living organisms as they breathe contaminated air, drink contaminated water, or eat contaminated food. These chemicals can become concentrated in the tissues and internal organs of animals and humans.

The concentration of acetone found in fish tissues is expected to be about the same as the average concentration of acetone in the water from which the fish was taken.

SUPPORT DOCUMENT: AQUIRE Database, ERL-Duluth, U.S. EPA, Phytotox.

Figure 3.2.1 EPA's Chemical Substance Database:
Acetone (7 of 7 pages)

 ToxFAQs

Acetone

September 1995

Agency for Toxic Substances and Disease Registry

This fact sheet answers the most frequently asked health questions about acetone. For more information, you may call 404-639-6000. This fact sheet is one in a series of summaries about hazardous substances and their health effects. This information is important because this substance may harm you. The effects of exposure to any hazardous substance depend on the dose, the duration, how you are exposed, personal traits and habits, and whether other chemicals are present.

> **SUMMARY:** Exposure to acetone results mostly from breathing air, drinking water, or coming in contact with products or soil that contain acetone. Exposure to moderate-to-high amounts of acetone can irritate your eyes and respiratory system, and make you dizzy. Very high exposure may cause you to lose consciousness. This chemical has been found in at least 572 of 1,416 National Priorities List sites identified by the Environmental Protection Agency.

What is acetone?
(Pronounced as'ah-tone)

Acetone is a manufactured chemical that is also found naturally in the environment. It is a colorless liquid with a distinct smell and taste. It evaporates easily, is flammable, and dissolves in water. It is also called dimethyl ketone, 2-propanone, and beta-ketopropane.

Acetone is used to make plastic, fibers, drugs, and other chemicals. It is also used to dissolve other substances.

It occurs naturally in plants, trees, volcanic gases, forest fires, and as a product of the breakdown of body fat. It is present in vehicle exhaust, tobacco smoke, and landfill sites. Industrial processes contribute more acetone to the environment than natural processes.

What happens to acetone when it enters the environment?

☐ A large percentage (97%) of the acetone released during its manufacture or use goes into the air.
☐ In air, about one-half of the total amount breaks down from sunlight or other chemicals every 22 days.
☐ It moves from the atmosphere into the water and soil by rain and snow. It also moves quickly from soil and water back to air.

Figure 3.2.2B Hazdat's ToxFAQ: Acetone (1 of 3 pages)

☐ Acetone doesn't bind to soil or build up in animals.
☐ It's broken down by microorganisms in soil and water.
☐ It can move into groundwater from spills or landfills.
☐ Acetone is broken down in water and soil, but the time required for this to happen varies.

How might I be exposed to acetone?

☐ Breathing low background levels in the environment
☐ Breathing higher levels of contaminated air in the workplace or from using products that contain acetone (for example, household chemicals, nail polish, and paint)
☐ Drinking water or eating food containing acetone
☐ Touching products containing acetone
☐ For children, eating soil at landfills or hazardous waste sites that contain acetone
☐ Smoking or breathing secondhand smoke.

How can acetone affect my health?

If you are exposed to acetone, it goes into your blood which then carries it to all the organs in your body. If it is a small amount, the liver breaks it down to chemicals that are not harmful and uses these chemicals to make energy for normal body functions. Breathing moderate-to-high levels of acetone for short periods of time, however, can cause nose, throat, lung, and eye irritation; headaches; light-headedness; confusion; increased pulse rate; effects on blood; nausea; vomiting; unconsciousness and possibly coma; and shortening of the menstrual cycle in women.

Swallowing very high levels of acetone can result in unconsciousness and damage to the skin in your mouth. Skin contact can result in irritation and damage to your skin.

The smell and respiratory irritation or burning eyes that occur from moderate levels are excellent warning signs that can help you avoid breathing damaging levels of acetone.

Health effects from long-term exposures are known mostly from animal studies. Kidney, liver, and nerve damage, increased birth defects, and lowered ability to reproduce (males only) occurred in animals exposed long-term. It is not known if people would have these same effects.

How likely is acetone to cause cancer?

The Department of Health and Human Services, the International Agency for Research on Cancer, and the Environmental Protection Agency (EPA) have not classified acetone for carcinogenicity.

Acetone does not cause skin cancer in animals when applied to the skin. We don't know if breathing or swallowing acetone for long periods will cause cancer. Studies of workers exposed to it found no significant risk of death from cancer.

Is there a medical test to show whether I've been exposed to acetone?

Methods are available to measure the amount of acetone in your breath, blood, and urine. The test can tell you how much acetone you were exposed to, although the amount that people have naturally in their bodies varies with each person. The tests can't tell you if you will experience any health effects from the exposure.

The test must be performed within 2-3 days after exposure because acetone leaves your body within a few days. These tests are not routinely performed at your doctor's office, but your doctor can take blood or urine samples and send them to a testing laboratory.

Figure 3.2.2B Hazdat's ToxFAQ: Acetone (2 of 3 pages)

Has the federal government made recommendations to protect human health?

The EPA requires that spills of 5,000 pounds or more of acetone be reported.

The Occupational Safety and Health Administration (OSHA) has set a maximum concentration limit in workplace air of 1,000 parts of acetone per million parts of air (1,000 ppm) for an 8-hour workday over a 40-hour week to protect workers.

The National Institute for Occupational Safety and Health (NIOSH) recommends an exposure limit of 250 ppm in workplace air for up to a 10-hour workday over a 40-hour workweek.

Glossary

Carcinogenicity: Ability to cause cancer
Evaporate: To change into a vapor or a gas
Ingesting: Taking food or drink into your body
Long-term: Lasting one year or longer

References

Agency for Toxic Substances and Disease Registry (ATSDR). 1994. Toxicological profile for acetone. Atlanta, GA: U.S. Department of Health and Human Services, Public Health Service.

Where can I get more information?

ATSDR can tell you where to find occupational and environmental health clinics. Their specialists can recognize, evaluate, and treat illnesses resulting from exposure to hazardous substances. You can also contact your community or state health or environmental quality department if you have any more questions or concerns.

For more information, contact:

Agency for Toxic Substances and Disease Registry
Division of Toxicology
1600 Clifton Road NE, Mailstop E-29
Atlanta, GA 30333
Phone: 404-639-6000
FAX: 404-639-6315

U.S. Department of Health and Human Services
Public Health Service
Agency for Toxic Substances and Disease Registry

Link to ATSDR Science Corner

Link to ATSDR Home Page

Charlie Xintaras / chx1@atsoaa1.em.cdc.gov

Figure 3.2.2B Hazdat's ToxFAQ: Acetone (3 of 3 pages)

MSDS for ACETONE   Page 1

---

1 - PRODUCT IDENTIFICATION

PRODUCT NAME: ACETONE
FORMULA: (CH3)2CO
FORMULA WT: 58.08
CAS NO.:   67-64-1
NIOSH/RTECS NO.:   AL3150000
COMMON SYNONYMS: DIMETHYL KETONE; METHYL KETONE; 2-PROPANONE
PRODUCT CODES: 9010,9006,9002,9254,9009,9001,9004,5356, A134,9007,
                       9005,9005,9008
EFFECTIVE: 08/27/86
REVISION #02

PRECAUTIONARY LABELLING

BAKER SAF-T-DATA(TM) SYSTEM

HEALTH	- 1	SLIGHT
FLAMMABILITY	- 3	SEVERE (FLAMMABLE)
REACTIVITY	- 2	MODERATE
CONTACT	- 1	SLIGHT

HAZARD RATINGS ARE 0 TO 4 (0 = NO HAZARD; 4 = EXTREME HAZARD).

LABORATORY PROTECTIVE EQUIPMENT

SAFETY GLASSES; LAB COAT; VENT HOOD; PROPER GLOVES; CLASS B
EXTINGUISHER

PRECAUTIONARY LABEL STATEMENTS

DANGER
CAUSES IRRITATION
EXTREMELY FLAMMABLE
HARMFUL IF SWALLOWED OR INHALED
KEEP AWAY FROM HEAT, SPARKS, FLAME. AVOID CONTACT WITH EYES, SKIN,
CLOTHING. AVOID BREATHING VAPOR. KEEP IN TIGHTLY CLOSED CONTAINER.
USE WITH ADEQUATE VENTILATION. WASH THOROUGHLY AFTER HANDLING.
IN CASE OF FIRE, USE ALCOHOL FOAM, DRY CHEMICAL, CARBON DIOXIDE -
WATER MAY BE INEFFECTIVE. FLUSH SPILL AREA WITH WATER SPRAY.

Figure 3.2.3 MSDS: Acetone (1 of 5 pages)

SAF-T-DATA(TM) STORAGE COLOR CODE: RED (FLAMMABLE)

---

2 - HAZARDOUS COMPONENTS

COMPONENT	%	CAS NO.
ACETONE	90-100	67-64-1

---

3 - PHYSICAL DATA

BOILING POINT: 56 C (133 F)   VAPOR PRESSURE(MM HG): 181

MSDS for   ACETONE   Page 2

MELTING POINT: -95 C (-139 F)   VAPOR DENSITY(AIR=1): 2.0

SPECIFIC GRAVITY: 0.79   EVAPORATION RATE: ~10
 (H2O=1)   (BUTYL ACETATE=1)

SOLUBILITY(H2O):   COMPLETE (IN ALL PROPORTIONS) % VOLATILES BY
                        VOLUME: 100

APPEARANCE & ODOR: CLEAR, COLORLESS LIQUID WITH A FRAGRANT SWEET
                        ODOR.

---

4 - FIRE AND EXPLOSION HAZARD DATA

FLASH POINT (CLOSED CUP: -18 C (0 F)   NFPA 704M RATING: 1-3-0

FLAMMABLE LIMITS: UPPER - 13.0 %   LOWER - 2.6 %

FIRE EXTINGUISHING MEDIA
 USE ALCOHOL FOAM, DRY CHEMICAL OR CARBON DIOXIDE.
 (WATER MAY BE INEFFECTIVE.)

SPECIAL FIRE-FIGHTING PROCEDURES
 FIREFIGHTERS SHOULD WEAR PROPER PROTECTIVE EQUIPMENT AND SELF-
 CONTAINED BREATHING APPARATUS WITH FULL FACEPIECE OPERATED
 IN POSITIVE PRESSURE MODE. MOVE CONTAINERS FROM FIRE AREA IF IT
 CAN BE DONE WITHOUT RISK. USE WATER TO KEEP FIRE-EXPOSED
 CONTAINERS COOL.

Figure 3.2.3 MSDS: Acetone (2 of 5 pages)

UNUSUAL FIRE & EXPLOSION HAZARDS
VAPORS MAY FLOW ALONG SURFACES TO DISTANT IGNITION SOURCES AND FLASH BACK.
CLOSED CONTAINERS EXPOSED TO HEAT MAY EXPLODE. CONTACT WITH STRONG OXIDIZERS MAY CAUSE FIRE.

---

5 - HEALTH HAZARD DATA

THRESHOLD LIMIT VALUE (TLV/TWA):   1780 MG/M3 (750 PPM)

SHORT-TERM EXPOSURE LIMIT (STEL):   2375 MG/M3 (1000 PPM)

PERMISSIBLE EXPOSURE LIMIT (PEL):   2400 MG/M3 (1000 PPM)

TOXICITY: LD50 (ORAL-RAT)(MG/KG)   - 9750
                 LD50 (ORAL-MOUSE)(MG/KG)   - 3000
                 LD50 (IPR-MOUSE)(MG/KG)   - 1297
                 LD50 (SKN-RABBIT) (G/KG)   - 20

CARCINOGENICITY: NTP: NO   IARC: NO   Z LIST: NO   OSHA REG: NO

EFFECTS OF OVEREXPOSURE
VAPORS MAY BE IRRITATING TO SKIN, EYES, NOSE AND THROAT.
INHALATION OF VAPORS MAY CAUSE NAUSEA, VOMITING, HEADACHE, OR LOSS OF CONSCIOUSNESS.
LIQUID MAY CAUSE PERMANENT EYE DAMAGE.
CONTACT WITH SKIN HAS A DEFATTING EFFECT, CAUSING DRYING AND IRRITATION.

MSDS for ACETONE   Page 3

INGESTION MAY CAUSE NAUSEA, VOMITING, HEADACHES, DIZZINESS, GASTROINTESTINAL IRRITATION. CHRONIC EFFECTS OF OVEREXPOSURE MAY INCLUDE KIDNEY AND/OR LIVER DAMAGE.

TARGET ORGANS
RESPIRATORY SYSTEM, SKIN

MEDICAL CONDITIONS GENERALLY AGGRAVATED BY EXPOSURE
NONE IDENTIFIED

ROUTES OF ENTRY
INHALATION, INGESTION, EYE CONTACT, SKIN CONTACT

Figure 3.2.3 MSDS: Acetone (3 of 5 pages)

EMERGENCY AND FIRST AID PROCEDURES
CALL A PHYSICIAN.
IF SWALLOWED, IF CONSCIOUS, IMMEDIATELY INDUCE VOMITING.
IF INHALED, REMOVE TO FRESH AIR. IF NOT BREATHING, GIVE ARTIFICIAL
RESPIRATION. IF BREATHING IS DIFFICULT, GIVE OXYGEN.
IN CASE OF CONTACT, IMMEDIATELY FLUSH EYES WITH PLENTY OF WATER
FOR AT LEAST 15 MINUTES. FLUSH SKIN WITH WATER.

---

6 - REACTIVITY DATA

STABILITY: STABLE   HAZARDOUS POLYMERIZATION: WILL NOT OCCUR

CONDITIONS TO AVOID: HEAT, FLAME, SOURCES OF IGNITION

INCOMPATIBLES: HALOGEN ACIDS AND HALOGEN COMPOUNDS, STRONG
                BASES, STRONG OXIDIZING AGENTS, CAUSTICS, AMINES AND
                AMMONIA, CHLORINE AND CHLORINE COMPOUNDS, STRONG
                ACIDS, ESP. SULFURIC, NITRIC, HYDROCHLORIC

---

7 - SPILL AND DISPOSAL PROCEDURES

STEPS TO BE TAKEN IN THE EVENT OF A SPILL OR DISCHARGE
    WEAR SUITABLE PROTECTIVE CLOTHING. SHUT OFF IGNITION SOURCES;
    NO FLARES, SMOKING, OR FLAMES IN AREA. STOP LEAK IF YOU CAN DO SO
    WITHOUT RISK. USE WATER SPRAY TO REDUCE VAPORS. TAKE UP WITH
    SAND OR OTHER NON-COMBUSTIBLE ABSORBENT MATERIAL AND PLACE
    INTO CONTAINER FOR LATER DISPOSAL. FLUSH AREA WITH WATER.

J. T. BAKER SOLUSORB(R) SOLVENT ADSORBENT IS RECOMMENDED FOR SPILLS
OF THIS PRODUCT.

DISPOSAL PROCEDURE
    DISPOSE IN ACCORDANCE WITH ALL APPLICABLE FEDERAL, STATE, AND
    LOCAL ENVIRONMENTAL REGULATIONS.

EPA HAZARDOUS WASTE NUMBER: U002 (TOXIC WASTE)

Figure 3.2.3 MSDS: Acetone (4 of 5 pages)

---

8 - PROTECTIVE EQUIPMENT

MSDS for   ACETONE   Page 4

VENTILATION: USE GENERAL OR LOCAL EXHAUST VENTILATION
                                TO MEET TLV REQUIREMENTS.

RESPIRATORY PROTECTION: RESPIRATORY PROTECTION REQUIRED IF
                                AIRBORNE CONCENTRATION EXCEEDS TLV. AT
                                CONCENTRATIONS UP TO 5000 PPM, A GAS MASK
                                WITH ORGANIC VAPOR CANNISTER IS
                                RECOMMENDED. ABOVE THIS LEVEL, A SELF-
                                CONTAINED BREATHING APPARATUS WITH FULL
                                FACE SHIELD IS ADVISED.

EYE/SKIN PROTECTION: SAFETY GLASSES WITH SIDESHIELDS, BUTYL
                                RUBBER GLOVES ARE RECOMMENDED.

---

9 - STORAGE AND HANDLING PRECAUTIONS

SAF-T-DATA(TM) STORAGE COLOR CODE: RED (FLAMMABLE)

SPECIAL PRECAUTIONS
    BOND AND GROUND CONTAINERS WHEN TRANSFERRING LIQUID. KEEP
    CONTAINER TIGHTLY CLOSED. STORE IN A COOL, DRY, WELL-VENTILATED,
    FLAMMABLE LIQUID STORAGE AREA.

---

10 - TRANSPORTATION DATA AND ADDITIONAL INFORMATION

DOMESTIC (D.O.T.)

PROPER SHIPPING NAME   ACETONE
HAZARD CLASS   FLAMMABLE LIQUID
UN/NA   UN1090
LABELS   FLAMMABLE LIQUID
REPORTABLE QUANTITY   5000 LBS.

INTERNATIONAL (I.M.O.)

PROPER SHIPPING NAME   ACETONE
HAZARD CLASS   3.1
UN/NA   UN1090
LABELS   FLAMMABLE LIQUID

Figure 3.2.3 MSDS: Acetone (5 of 5 pages)

TR-2 Carcinogenesis Bioassay of Trichloroethylene (CAS No. 79-01-6)

Trichloroethylene (TCE), a halogenated chemical, has been tested for carcinogenicity in the National Cancer Institute's Carcinogenesis Bioassay Program. Trichloroethylene has been used primarily as a solvent in industrial degreasing operations. Other uses have been as a solvent in dry cleaning and food processing, as an ingredient in printing inks, paints. etc., and as a general anesthetic or analgesic.

Industrial grade (>99% pure) trichloroethylene was tested using 50 animals per group at 2 doses and with both sexes of Osborne-Mendel rats and B6C3F1 mice. Twenty of each sex and species were maintained as matched controls, in addition to colony and positive carcinogen controls. Animals were exposed to the compound by oral gavage 5 times per week for 78 weeks. At the end of treatment, animals were observed until terminal sacrifice at 110 weeks for rats and 90 weeks for mice. A complete necropsy and microscopic evaluation of all animals (except 7 of the original 480) was conducted.

Two doses were used with animals started on test at approximately 6 weeks of age. The initial doses used in this test were the estimated maximum tolerated dose (MTD) and 1/2 MTD, as predicted from data obtained in a 6-week toxicity study. For rats, the initial doses were 1,300 and 650 mg/kg body weight. These were changed, based upon survival and body weight data, so that the "time-weighted average" doses were 549 and 1,097 mg/kg for both male and female rats. For mice, the initial doses were 1,000 and 2,000 mg/kg for males and 700 and 1,400 mg/kg for females. The doses were increased so that the "time-weighted average" doses were 1,169 and 2,339 mg/kg for male mice and 869 and 1,739 mg/kg for female mice.

Clinical signs of toxicity, including reduction in weight, were evident in treated rats. These, along with an increased mortality rate necessitated a reduction in doses during the test. In contrast, very little evidence of toxicity was seen in mice, so doses were increased slightly during the study. The increased mortality in treated male mice appears related to the presence of liver tumors.

A variety of neoplastic lesions were observed in rats with no significant difference between trichloroethylene-treated and control animals. The only lesion that might be attributed to the treatment was a chronic nephropathy found in both sexes and at both dose levels.

With both male and female mice, primary malignant tumors of the liver, i.e., hepatocellular carcinoma, were observed in high numbers. For males, 26/50 low dose and 31/48 high dose animals had hepatocellular carcinomas as compared with 1/20 matched controls and 5/77 colony controls. The differences between treated and matched control males at both doses were highly significant (P<0.01). For females, hepatocellular carcinomas were observed in 4/50 low dose and 11/47 high dose animals as compared with 0/20 matched controls and 1/80 colony controls. While the difference between the high dose female mice and matched controls was also highly significant (P<0.01), the difference at the low dose was less (P=0.09). For both male and female mice, age-adjusted tests for linear trend (dose response) were highly significant for hepatocellular carcinoma (P<0.001 for males and P=0.002 for females).

In male mice at the high doses, hepatocellular carcinomas were observed early in the study. The first was seen at 27 weeks; 9 others were found in male mice dying by the 78th week. The tumor was not observed so early in low dose male or female mice. The diagnosis of hepatocellular

---

Figure 3.2.4A Long-Term Toxicological Study Abstract:
Trichloroethylene (1 of 2 pages)

carcinoma was based on size, histologic appearance, and presence of metastasis, especially to the lung. No other lesion was significantly elevated (P<0.05) in treated mice. The incidence of hepatocellular carcinomas in the trichloroethylene-matched controls was typical of that observed in colony controls.

Carbon tetrachloride (CCl4) was used as a positive control for the series of chlorinated chemicals which included trichloroethylene. While virtually all male and female mice developed hepatocellular carcinomas following carbon tetrachloride treatment, the response in the Osborne-Mendel rats was considerably less. Only about 5% developed hepatocellular carcinomas. Thus, there appears to be a marked difference in sensitivity to induction of carcinomas by chlorinated compounds between the B6C3F1 mouse and the Osborne-Mendel rat.

The results of this carcinogenesis test of trichloroethylene clearly indicate that trichloroethylene induced a hepatocellular carcinoma response in mice. While the absence of a similar effect in rats appears most likely attributable to a difference in sensitivity between the Osborne-Mendel rat and the B6C3F1 mouse, the early mortality of rats due to toxicity must also be considered.

Synonyms: trichloroethene; acetylene trichloride; ethinyl trichloride; 1,1,2-trichloroethylene, TCE

Report Date: February 1976
Levels of Evidence of Carcinogenicity:

Male Rats:	Negative
Female Rats:	Negative
Male Mice:	Positive
Female Mice:	Positive

Note: Trichloroethylene was subsequently studied by gavage in F344 rats and B6C3F1 mice (See TR-243, reported in 1990) and also in four strains of rats (ACI, August, Marshall, and Osborne-Mendel) by gavage (See TR-273, reported in 1988)

Figure 3.2.4A Long-Term Toxicological Study Abstract:
Trichloroethylene (2 of 2 pages)

TOX-3 Toxicity Studies of Acetone in F344/N Rats and B6C3F1 Mice (Drinking Water Studies) (CAS No. 67-64-1)

Toxicity studies were conducted by administering acetone (greater than 99% pure) in drinking water to groups of F344/N rats and B6C3F1 mice of each sex for 14 days or 13 weeks.

Fourteen-Day Studies: All rats and mice receiving concentrations as high as 100,000 ppm acetone in drinking water lived to the end of the 14-day studies. The mean body weights of male rats receiving 50,000 or 100,000 ppm and female rats given 100,000 ppm were lower than those of controls. Body weights of all groups of mice were similar. Kidney and liver weight to body weight ratios for exposed rats and mice were greater than those for controls. Histopathologic changes were not seen in these organs in rats or in the kidney in mice. Centrilobular hepatocellular hypertrophy was noted in male and female mice receiving 20,000 and 50,000 ppm acetone, respectively.

Thirteen-Week Studies: All rats lived to the end of the 13-week studies (drinking water concentrations as high as 50,000 ppm). The final mean body weights of rats receiving 50,000 ppm were 19% lower than that of controls for males and 7% lower for females. Water consumption by all rats that received 50,000 ppm acetone and females that received 20,000 ppm or more was notably lower than that by controls. Liver and kidney weight to body weight ratios were increased for male and female rats receiving 20,000 ppm or greater. Caudal and right epididymal weights and sperm motility were decreased for male rats given 50,000 ppm, and the percentage of abnormal sperm was increased. Leukocytosis and thrombocytopenia were observed at 20,000 ppm and above (males and females), and reticulocytopenia and erythrocytopenia were seen at 5,000 ppm and above (males). These changes, in addition to increase in erythrocyte size (MCV), are consistent with macrocytic anemia. Splenic pigmentation (hemosiderosis) noted in dosed male rats was apparently related to these changes. The increased incidence and severity of nephropathy observed in dosed male rats were considered the most prominent chemically related findings in this study.

All mice lived to the end of the 13-week studies (drinking water concentrations up to 20,000 ppm for males and up to 50,000 ppm for females). The final mean body weights of dosed and control mice were similar. Water consumption by female mice that received 50,000 ppm acetone was notably lower than that by controls. The absolute liver weight and the liver weight to body weight ratio were significantly increased for females receiving 50,000 ppm, and the absolute spleen weight and the spleen weight to body weight ratio were significantly decreased. Results from the hematologic analyses did not show any biologically significant effects. Centrilobular hepatocellular hypertrophy of minimal severity was seen in 2110 female mice receiving 50,000 ppm. No compound-related lesions were found in male mice.

In summary, the results from these studies show that acetone is mildly toxic to rats and mice when administered in drinking water for 13 weeks. Minimal toxic doses were estimated to be 20,000 ppm acetone for male rats and male mice and 50,000 ppm acetone for female mice. No toxic effects were identified for female rats. The testis, kidney, and hematopoietic system were identified as target organs in male rats, and the liver was the target organ for male and female mice.

Synonyms: 2-propanone; dimethyl ketone; pyroacetic acid

Report Date: January 1991

Figure 3.2.4B Short-term Toxicalogical Study Abstract:
Acetone (1 of 1 page)

NTP Report on the IMMUNOTOXICITY OF GALLIUM ARSENIDE (CAS No. 1303-00-0) in Female B6C3F1 Mice (immune modulation studies)

Introduction

Gallium arsenide is used extensively in the semiconductor industry for the manufacture of various electronic components including light-emitting diodes. In light of studies suggesting that arsenic compounds are immunotoxic, the potential effects of gallium arsenide on the immune system of mice were studied following intratracheal exposure.

Design:

Gallium arsenide (Lot #M100386/B04) was obtained from Alpha Products (Danvers, MA.) and was determined by HPLC to be > 98% pure. The material was ground in a mortar and pestle and administered to female B6C3F1 mice as a suspension of particles in saline containing 0.05% Tween 80; mean particle size of 1.5 mm. Prior to exposure, mice were anesthetized by an intravenous injection of hexobarbital (80 mg/kg). Mice received a single intratracheal instillation of the material at dose levels of 50, 100 or 200 mg/kg in a volume of 0.1ml. Control groups received vehicle or 10 mg/kg sodium arsenite (Sigma Chemical Corp.) also by the intratracheal route. Animals were sacrificed 15 days after the exposures for immune testing. For each of the parameters examined, usually 8 mice were examined per treatment and vehicle group except the sodium arsenite control group which consisted of 4 mice per group.

Results:

There were no treatment-related effects on mortality. In most cases there were no effects on body weight although in one replicate experiment a decrease in weight was observed in the high dose group and body weight gain was decreased over the 2 week post-exposure. Spleen weights were increased in all gallium arsenide treated groups while thymus weights were decreased at the high dose (Table 1). A dose-response increase in lung weights occurred following treatment but none of the individual values were significant. The lungs were characterized by darkened areas consisting of lymphocyte and macrophage infiltration. There was a tendency (not significant) for a treatment-related decrease in peripheral leukocyte counts (Table 1) which was associated with a significant decrease in peripheral lymphocyte counts in the 200 mg/kg treatment group (not shown).

The immune and bone marrow studies are also summarized in Table 1. Gallium arsenide suppressed the following immune parameters dose- dependently: the IgM and IgG (not shown) antibody response to sheep erythrocytes, the delayed hypersensitivity response to KLH, the mixed leukocyte response (MLR), and, to a lesser extent, splenic B lymphocyte numbers. There were no effects on mitogenesis or T lymphocyte numbers. Although there was a treatment-related increase in the natural killer cell response, the values from the control group were below historical control values. Bone marrow changes were not observed except for an increase in CFU-C1s when based upon total cellularity (data not shown).

Table 2 summarizes the host resistance studies in mice exposed to gallium arsenide. Exposure to gallium arsenide increased the resistance to Listeria monocytogenes and to a lesser extent, PYB6 tumor cell growth, while decreasing resistance to pulmonary burden of B16F10 melonoma. There were no consistent treatment-related effects on resistance to Streptococcus pneumoniae.

Figure 3.2.4C Immunotoxicity Study Abstract:
Gallium Arsenide (1 of 3 pages)

Conclusion:

Under these experimental conditions, gallium arsenide produced multiple immunotoxic effects. The effects could result from a direct action of the chemical on lymphoid cells or indirectly via local pulmonary damage. The later may account for the variability observed in the host resistance models.

Table 1: Summary of Gallium Arsenide Immune Studies

Parameter	0	50	100	200	sodium arsenite 10 mg/kg
		Dose (mg/kg)			
Body Weight (g)	22.0	21.6	21.7	20.9	21.9
Spleen (mg)	72	81*	89#	103#	81
Spleen:Body wt. (%)	0.33	0.38#	0.41#	0.50#	0.37#
Thymus (mg)	61	49	60	42*	60
Thymus:Body wt. (%)	0.28	0.22	0.28	0.20*	0.28
Lung (mg)	425	480	533	551	449
Lung: Body Wt (%)	1.95	2.23	2.48	2.69	2.27
Peripheral Leukocyte Count($10E3/mm3$)	4.0	3.2	2.9	2.8	3.7
Spleen cell no. (x 10E7)	16.8	14.7	13.2#	12.0#	15.1
Thy 1.2+(%)	51	47	48	46	52
sIg+ (%)	56	53*	49*	46#	52#
Delayed Hypersensitivity Response (SI)	4.0	3.1	2.3	1.1#	2.6
NK Cell Activity (% cytotoxicity)	7.4	14.5#	16.3#	14.9#	11.6*
IgM PFC/106 cells	2013	1432*	1227#	694#	1529*
IgM PFC/spleen (x 10E3)	340	206#	162#	84#	228#
Lymphoproliferation 3H-TdR incorporation (cpm x 10E3):					
Concanavalin A	126	94	128	105	100
Phytohemagglutinin	136	109	151	125	104
Lipopolysaccharide	67	59	89	74	55
Mixed Leukocyte Resp.	28	11#	11#	10#	29

Figure 3.2.4C Immunotoxicity Study Abstract:
Gallium Arsenide (2 of 3 pages)

Parameter	Dose (mg/kg)				sodium arsenite
	0	50	100	200	10 mg/kg
Bone Marrow					
Cells/Femur (x10E6)	9.7	10.6	11.4	11.7	12.3
DNA Synthesis					
(cpm x 10E3)	58	60	58	61	73
CFU-C1/105					
Nucleated Cells^	99	107	116	111	95
CFU-C2/105					
Nucleated Cells^	59	55	59	52	59

note:

^ CFU-C1 and C2 are defined in Appendix A.
* different from vehicle control at $P < 0.05$
# different from vehicle control at $P < 0.01$

Table 2. Summary of Gallium Arsenide Host Resistance Studies

Host Resistance Test	Dose mg/kg (No. Per Group)				Sodium arsenite, (10 mg/kg)
	0	50	100	200	
L. Monocytogenes	(12)	(12)	(12)	(12)	(8)
% Morbidity	67	58	8#	8#	0
S pneumoniae					
% Morbidity	(12)	(12)	(12)	(12)	(11)
Low Infective					
Challenge	17	0	25	17	0
% Morbidity	(12)	(12)	(12)	(12)	(11)
High Infectivity					
Challenge	82	58	67	75	91
PYB6 Tumor- % pos.	(12)	(12)	(12)	(12)	(12)
21 days post transplant	33	58	17	17	8
B16F10 Melanoma	(16)	(16)	(16)	(15)	(16)
CPM (x 10E3)/Lung	3,9	5,1	8.0#	22.0#	4.0

Figure 3.2.4C Immunotoxicity Study Abstract:
Gallium Arsenide (3 of 3 pages)

Pesticide/Fertilizer Contamination-Mixture II CAS No. PESTFERTMIX2
RACB90027
NTIS # PB92-140730/AS

Abstract

Pesticide/Fertilizer Mixture II (PFM II) is composed of six pesticides (Aldicarb, Atrazine, Dibromochloropropane, 1,2- Dichloropropane, Ethylene Dibromide, and Simazine) and 1 fertilizer (Ammonium nitrate) and represents a real-life mixture of ground water contamination in California. PFM II in drinking water was tested for its effects on fertility and reproduction in Swiss CD-1 mice according to the Continuous Breeding Protocol. 0, 1, 10, and 100X (1X being the median concentration found in ground water) were chosen to investigate the effects on fertility and reproduction.

Male and female mice (F0) were continuously exposed for a 7-day precohabitation and a 98-day cohabitation period (Task 2). Male and female body weights in Task 2 were within 10% of control values. Water consumption was similar in F0 control and treated animals. Fertility and reproduction in F0 animals were not effected by PFM II.

The F1 pups from the final litter in the control, 10 and 100X groups were weaned for second generation studies. At necropsy, kidney/adrenal weights were approximately 10% higher than controls in F1 10X-treated males and females and 100X-treated females. Absolute seminal vesicle weight as well as seminal vesicle-to-body weight ratio significantly decreased (ca. 11%) in the 100X dose group. PFM II produced mild systemic toxicity and no reproductive toxicity in Swiss CD-1 mice at the dose levels tested.

Figure 3.2.4D Reproductive Toxicology Study Abstract:
Pesticide/Fertilizer Mixture (1 of 1 page)

Acetone (CAS # 67-64-1) (Sprague-Dawley Rats & Swiss (CD-1®) Mice)

Abstract

Acetone, an aliphatic ketone, is a ubiquitous industrial solvent and chemical intermediate; human exposure potential is high. Acetone was evaluated for developmental toxicity through inhalation in both Sprague-Dawley rats at 0, 440, 2,200, or 11,000 ppm, and in Swiss (CD-1®) mice at 0, 440, 2200, and 6600 ppm, for 6 hr/day, 7 days/week. Each treatment group consisted of 10 non- pregnant virgin females, and 32 positively mated rats or mice. Positively mated mice were exposed on days 6-17 of gestation (gd), and rats on 6-19 gd. The day of plug or sperm detection was designated as gd 0. Body weights were obtained throughout the study period, and uterine and fetal body weights were obtained at sacrifice (rats, 20 gd; mice, 18 gd). Implants were counted and their status recorded. Live fetuses were sexed and examined for gross, visceral, skeletal, and soft-tissue craniofacial defects.

Pregnant rats did not exhibit overt symptoms of toxicity other than significant differences for the 11,000 ppm group in body weight (14, 17, 20 gd), reduced cumulative weight gain from 14 gd onward, uterine weight and in extragestational weight gain. There were no maternal deaths, and organ weights remained unchanged. Also unchanged were the number of implantations, the mean percent of live pups/litter, the mean percent of resorptions/litter, or the fetal sex ratio. However, fetal weights were significantly reduced for the 11,000 ppm group by 7.5%, and there were more cases of diverse malformations (though still not significant) in this group. There was no increase in the incidence of fetal variations, reduced ossification sites, or in the mean incidence of fetal variations per litter.

Swiss (CD-1®) mice exhibited severe narcosis at 11,000 ppm acetone; consequently, the high exposure concentration was reduced to 6,600 ppm acetone after one day of exposure. There was no toxicity nor maternal death at any dose. There was no effect on any reproductive index, or on the fetal sex ratio. Developmental toxicity was observed in mice in the 6600 ppm exposure group as: 1) a 7.5% reduction in fetal weight, and 2) a slight increase (from 3.2 to 7.8%) in the percent incidence of late resorptions. This increase in late resorptions was not sufficient to cause a decrease in the mean number of live fetuses per litter.

The 2200 ppm acetone level was the no observable effect level (NOEL) in both the Sprague-Dawley (CD) rat and the Swiss (CD-1®) mouse for developmental toxicity. However, acetone was a developmental toxicant at 11,000 ppm in rats, and at 6,600 ppm in mice; these doses were not maternally toxic.

Figure 3.2.4E Teratology Study Abstract: Acetone (1 of 1 page)

CAL/EPA ACCESS

LIBRARY: DTSC

LIBRARY FILE NAME: PESTICI

DOCUMENT ISSUE DATE:

PESTICIDE RINSATES: Biodegradation Technology (January 1994)

In the past, when government regulation on pesticide wastes was minimal, pesticide applicators would wash down their aircraft and ground rigs with large amounts of water and let the contaminated water run onto the ground or into ditches. Eventually this practice not only resulted in contaminated soil, but some of the hazardous constituents leached into the ground waters below or ran off into surface waters. With increased government regulation, the applicators were required to collect the exterior rinse waters and treat them as hazardous waste. These wastes are typically stored until there is enough to ship to a Class I landfill or incinerate. Landfilling includes chemical stabilization to reduce mobility and toxicity of the hazardous constituents before treatment by solar evaporation or cyanide oxidation.

The interior tank rinsates are not a waste problem as they may be collected and either used to make the next batch of pesticides or field rinsed. Field rinsing is the application of the product in a diluted form to the field for the chemical's intended purpose, according to label directions.

Because of the high cost and liability involved with disposing of pesticide waste legally, there are still many cases of illegally disposed pesticides each year. If the responsible parties can be found they may not only be fined thousands of dollars, but may also face criminal charges. But whether these violators are identified or not, the cost to tax payers to clean up the sites, along with increased environmental problems, is significant.

In California during 1990, nearly 182 million pounds of pesticides were used by over 4,500 licensed pesticide applicators in agriculture and forestry. The total number of jobs it took to apply these pesticides was over 2 million. Assuming that for every job 50 gallons of water (which is a conservative estimate) was used to wash down the exterior of the application equipment, there were potentially 100 million gallons of hazardous rinsate generated and subsequently landfilled, incinerated or illegally dumped.

The Advanced Bio-Gest (ABG) system offers an alternative to disposal and incineration.

DESCRIPTION

Biodegradation of organic matter is a natural cycle. Every naturally occurring organic compound can be decomposed by one or more organisms. Because microorganisms have high metabolic potential, coupled with very fast reproduction, they are able to rapidly convert naturally occurring complex organic matter into carbon dioxide, water and harmless salts. The ABG system applies this principle in a new way.

Figure 3.2.8 California's DTSC Information:
Pesticide Rinsates (1 of 3 pages)

The ABG system is a biologically based, electro mechanical system that uses horse manure as its source of microbes. Horse manure seems to be the best source for these microbes since the horse uses the same process to detoxify poisonous plants in its own stomach. Unlike most other herbivores, horses do not have a harsh sterilizing environment in the last part of their intestines, so the microbes pass through to the manure. The manure is placed in a stainless steel chamber in the ABG system and acts as a digestive filter medium. Toxic organic waterborne wastes are pumped and poured onto the digesting medium. To assure the proper rate of digestion, the system is maintained in an aerobic condition by blowing heated air through the digesting filter medium.

The warm air also picks up moisture and any airborne products produced from the digestion of toxic waste. The moisture laden warm air passes over a grid of condenser coils. The water vapor condenses into demineralized water and is used for pesticide make up water or for wash down water. The efficiency rate for the recycled water that is produced is approximately 30%. The University of California, Davis, has also tested a condensing system which recycles up to 90% of the water.

The overall dimensions of the ABG system are approximately 5' x 3' x 4.5'. The digestion chamber holds five cubic feet of manure.

The maintenance required for the ABG system is simply to fill the influent tank with waste water. The flow of waste in, and recyclable water out, can be set up for automatic continuous feed. The manure acts as a disappearing filter and may need to be replenished at the rate of approximately one half to one cubic foot per month. The medium needs to be replaced only if the salts that remain reach levels that are harmful to the microbes. The activated charcoal regenerates itself due to bacterial growth transferred from the digestion media and it is not anticipated that this will need replacement.

ECONOMICS

The cost of landfilling pesticide rinsates is approximately $14.91 per gallon, and incineration is $10.06 per gallon of waste. The capital cost of the ABG system is $14,950. The average treatment cost using this system is $1.15 per gallon of pesticide rinsate. The ABG system is much more attractive economically than either land disposal or incineration.

The ABG system also eliminates the need for transportation and the liabilities associated with it. The closed loop system can be added for a cost of $2,485.

PUBLICATIONS

A hard copy of this fact sheet (Document Number 1216) is available upon request. It contains a schematic of the Advanced Bio-Gest Aerobic Prototype System and two tables listing related costs. In addition, the following types of reports are available from the Department of Toxic Substances Control: final reports summarizing this and other hazardous waste reduction grant projects, waste audit studies of specific industries, waste reduction alternatives for specific waste streams, case studies of alternative technologies, and economic implications of using waste reduction methods.

Figure 3.2.8 California's DTSC Information:
Pesticide Rinsates (2 of 3 pages)

Please contact the Technology Development Branch at:

Department of Toxic Substances Control
Office of Pollution Prevention and Technology Development
P.O. Box 806
Sacramento, CA 95812-0806
(916) 322-3670

This system was fabricated in Willits, California, by Advanced Manufacturing and Development, Inc., and tested at the University of California, Davis. The mention of a trade name, commercial product, or organization does not constitute endorsement or recommendation for use by the Department of Toxic Substances Control.

Figure 3.2.8 California DTSC Information: Pesticide Rinsates

National Water Conditions

United States Department of the Interior Geological Survey

STATION NAME: OHIO RIVER AT DAM 53 NEAR GRAND CHAIN
STATION NUMBER: 03612500
STATE: ILLINOIS
COUNTY: BOONE

STATISTICAL SUMMARY OF SELECTED WATER QUALITY DATA
COLLECTED FROM OCT 1954 TO JUN 1994

DESCRIPTIVE STATISTICS

#WATER-QUALITY CONSTITUENT	SAMPLE SIZE	MAXIMUM	MINIMUM	MEAN	9
00061 Discharge, Inst. CFS	168	1060000	36900.000	290419.625	76059
00095 Specific Conduct US/CM @ 25c	749	693.000	150.000	328.024	47
00010 Water Temperature (Degrees)	397	31.000	0.500	17.105	2
00400 PH, WH, FIELD (Standard Unit)	682	9.000	5.500	7.395	
00410 Alkalinity, WH, FE (MG/L AS CACO3)	350	120.000	39.000	74.474	10
00418 Alkalinity, DIS, F (MG/L AS CACO3)	39	102.000	53.000	75.692	9
00453 Bicarbonate, DIS, (MG/L AS HCO3)	35	124.000	65.000	91.200	11
39086 Alkalinity, DIS,I (MG/L AS CACO3)	35	102.000	53.000	74.743	9
00300 Oxygen Dissolved (MG/L)	141	15.000	3.100	9.185	1
00301 Oxygen Dis. Perc. % of Saturatio	114	125.000	6.000	88.439	10
BACTERIA					
31625 Coliform Fecal 0 Cols. /100 ML	115	2000.000	1.000	143.991	53
31673 Fecal Strpt KF A Cols. 100 ML	120	6400.000	3.000	403.500	169
NUTRIENTS					
00631 NO2 + NO3 Dissol (MG/L AS N)	165	2.800	0.190	1.005	
00608 Nitrogen Ammonia (MG/L AS N)	161	0.240	<0.010	0.057*	
00613 Nitrogen, Nitrite (MG/L AS N)	116	0.130	<0.010	0.026*	
00671 Phosphorus Ortho (MG/L AS P)	150	0.220	<0.010	0.037*	
00666 Phosphorus Diss. (MG/L AS P)	113	0.340	<0.010	0.047*	
00625 Nitrogen A+O TOT.(MG/L AS N)	240	2.400	0.040	0.608	
00665 Phosphorus Total (MG/L AS P)	240	1.000	0.020	0.139	

CFS	= CUBIC FEET PER SECOND	A+O	= AMMONIA PLUS ORGANIC
US/CM @ 25C	= MICROSEIMENS PER CENTIMETER AT 25 DEGREES CELSIUS	N	= NITROGEN
MG/L	= MILLIGRAMS PER LITER	NO2	= NITRITE
CACO3	= CALCIUM CARBONATE	NO3	= NITRATE
HCO3	= BICARBONATE	P	= PHOSPHORUS
COLS./100 ML	= COLONIES PER 100 MILLILITER		
%	= PERCENT		
<	= LESS THAN		

Unless otherwise noted, all nutrients results are from filtered samples.

*VALUE IS ESTIMATED BY USING A LOG-PROBABILITY REGRESSION
#TO SEE COMPLETE NAME OF CONSTITUENTS, CLICK ON HIGHLIGHTED TEXT ABOVE

Figure 3.3.5A National Water Conditions' Water Quality Information, Ohio River (1 of 1 page)

National Water Conditions

United States Department of the Interior Geological Survey

Provisional data; subject to revision

**WATER LEVELS IN KEY OBSERVATION WELLS IN SOME REPRESENTATIVE AQUIFERS
IN THE CONTERMINOUS UNITED STATES—SEPTEMBER 1995**

Ground-Water Region Aquifer and Location	Depth of well in feet	Water level in feet below land-surface datum	Departure from average in feet	Net change in water level in feet since: Last month	Last Year
WESTERN MOUNTAIN RANGES (1)					
Rathdrum Prairie aquifer near Spirit Lake, Idaho	448	411.1	.5	-.1	8.1
ALLUVIAL BASINS (2)					
Alluvial valley-fill aquifer in Steptoe Valley, Nevada	122	10.12	2.45	-.05	.07
Valley-fill aquifer near Tucson, Arizona	542	376.06	-16.81	-.91	-5.62
Hueco bolson aquifer at El Paso, Texas	640	279.07	-22.31	.53	-2.98
COLUMBIA LAVA PLATEAU (3)					
Snake River Plain aquifer near Eden, Idaho	137	127.5	-11.3	.7	-3.6
Columbia River basalts aquifer at Pendleton, Oregon	1,501	236.30	-38.37	.74	-1.35
COLORADO PLATEAU AND WYOMING BASIN (4)					
Dakota aquifer near Blanding, Utah	140	42.57	2.66	.10	.21
HIGH PLAINS (5)					
Ogallala aquifer near Colby, Kansas	175	131.67	-10.37	-.27	-.26
Southern High Plains aquifer at Lovington, New Mexico	212	58.49	-3.52	.07	-.44
NONGLACIATED CENTRAL REGION (6)					
Sentinel Butte aquifer near Dickinson, North Dakota	160	...	...	...	...
Sand and gravel Pleistocene aquifer near Valley Center Kansas	54	17.66	-.20	-.40	2.43
Glacial outwash aquifer near Louisville, Kentucky	94	18.51	5.31	-.18	-.61
Conemaugh Group aquifer near Halleck, West Virginia	141	82.40	-8.19	-.45	-3.86
GLACIATED CENTRAL REGION (7)					
Fluvial sand and gravel aquifer, Platte River Valley, near Ashland, Nebraska	12	6.07	.16	-.54	-.66
Sheyenne Delta aquifer near Wyndmere, North Dakota	40	...	...	...	...
Shallow drift aquifer near Roscommon, Michigan	14	5.21	-.19	-.48	-.63
Silurian-Devonian carbonate aquifer near Dola, Ohio	51	9.11	.06	-1.57	-1.11
PIEDMONT AND BLUE RIDGE (8)					
Water-table aquifer in Petersburg Granite, southeastern Piedmont at Colonial Heights, Virginia	100	17.12	-.95	-.19	-1.53
Weathered granite aquifer near Mocksville, North Carolina	31	15.00	3.10	.23	.41
Surficial aquifer at Griffin, Georgia	30	...	...	...	...
NORTHEAST AND SUPERIOR UPLANDS (9)					
Pleistocene glacial outwash aquifer, at Camp Ripley, near Little Falls, Minnesota	59	...	...	...	...
Glacial outwash sand aquifer at Oxford, Maine	39	...	...	...	...
Shallow sand aquifer (glacial deposits) at Acton, Massachusetts	34	20.58	-.70	-.55	-.74
Stratified drift aquifer near Morristown, Vermont	50	20.19	-.25	-.46	-.24

Figure 3.3.5B National Water Conditions'
Aquifer Information (1 of 2 pages)

Ground-Water Region Aquifer and Location	Depth of well in feet	Water level in feet below land-surface datum	Departure from average in feet	Net change in water level in feet since: Last month	Last Year
ATLANTIC AND GULF COASTAL PLAIN (10)					
Columbia deposits aquifer near Camden, Delaware	11	9.26	-1.91	-.39	-2.49
Memphis sand aquifer near Memphis, Tennessee	384	109.88	-18.96	-.23	-1.34
Eutaw aquifer at Montgomery, Alabama	270	29.7	-5.0	.5	-.6
Evangeline aquifer at Houston, Texas	1,152	270.33	32.76	-2.26	9.26
SOUTHEAST COASTAL PLAIN (11)					
Upper Floridan aquifer on Cockspur Island, near Savannah, Georgia	348	...	...	...	...
Upper Floridan aquifer at Jacksonville, Florida	905	-23.4	-4.3	.6	.4

[Return to Table of Contents]

Figure 3.3.5B National Water Conditions'
Aquifer Information (2 of 2 pages)

Gopher Menu

■ Search Clean Water Act / Safe Drinking Water Act data
■ 12/19/94 - Revisions to Drinking Water Analytical Methods
■ 7/12/94 - Revisions to National Primary Drinking Water Regulations
■ 02/03/94 - Indian Tribes as States for Purposes of CWA
■ 12/29/93 - SDWA Regulation Clarifying Requirements for Certain UIC Wells
■ 11/22/93 - The Environmental Protection Agencys Policy and Technical Guidance
■ 10/8/93 - Clean Water Act Regulatory Programs - Final Rule and Related
■ 11/23/92 - Developing Pollution Prevention Plans and Best Management
■ 9/10/92 - Final NPDES General Permits for Storm Water Discharges Associated
■ 8/27/92 - National Pollutant Discharge Elimination System Application
■ 8/26/92 - Announcement of DOE Drinking Water Systems Workshop
■ 7/30/92 - Final Drinking Water Regulation for 23 Contaminants
■ 7/14/92 - Corrections to National Primary Drinking Water Regulations
■ 6/22/92 - Review of the Environmental Protection Agencys Interim Guidance
■ 6/18/92 - Status of Compliance with Federal Storm Water Regulations
■ 4/22/92 - Clean Water Act (CWA) (Section 404)
■ 3/12/92 - Review of Planned USGS National Water Information System II
■ 12/4/91 - Request for Review of Notice on Phase V Drinking Water Chemicals
■ 12/4/91 - Request for Information on State Wastewater Permits for Discharg
■ 10/11/91 - Workshop on the National Pollutant Discharge Elimination System
■ 8/23/91 - National Pollutant Discharge Elimination System General Permits
■ 8/23/91 - 1989 - Federal Manual for Identifying and Delineating Jurisdicti
■ 8/15/91 - The National Pollutant Discharge Elimination System Permit
■ 7/31/91 - Proposed Drinking Water Standards for Radionuclides
■ 7/22/91 - Drinking Water Standards for Lead and Copper
■ 7/22/91 - Drinking Water Standards Final Rule
■ 4/16/91 - Proposal To Amend the Nationwide Permit Program Regulations
■ 4/16/91 - Additional Guidance on the National Pollutant Discharge
■ 9/19/90 - Safe Drinking Water Act - Proposed Rule on the National Primary
■ 5/18/90 - List of Outstanding Notices of Violation and Warning Letters
■ 6/21/89 - Safe Drinking Water Act, National Primary and Secondary Drinking
■ 2/7/89 - Proposed Rule for the National Pollutant Discharge Elimination
■ 9/23/88 - ENVIRONMENTAL GUIDANCE ADVISORY - Proposed Rule Maximum
Contaminant Level
■ 8/25/88 - ENVIRONMENTAL GUIDANCE ADVISORY
■ 7/14/88 - Corrections to the National Primary Drinking Water Regulations
■ 5/5/88 - ENVIRONMENTAL GUIDANCE ADVISORY - Safe Drinking Water Act
Amendme
■ 4/27/88 - Los Alamos National Laboratory NPDES Permit
■ 2/26/88 - ENVIRONMENTAL GUIDANCE ADVISORY - Environmental Protection
Agency
■ 12/8/86 - Safe Drinking Water Act Amendment - Prohibition on the Use of Lead
■ 4/24/86 - Water Quality Limits for Radioactivity in Surface Water

---

Figure 3.3.9 Environmental Guidance Documents for the
Clean Water Act and Groundwater (1 of 2 pages)

Gopher Menu

- Search Groundwater data
- 10/14/94 - DOE Ground Water Protection Work Group Meeting
- 5/6/94 - DOE GROUND-WATER PROTECTION WORK GROUP
- 12/06/93 - Technical Impracticability of Ground Water Remediation in ER
- 8/11/92 - Groundwater Remediation Considerations in Environmental
- 6/10/92 - Department of Energy Comments on EPAs Draft Comprehensive State
- 5/22/92 - Interagency Workshop on Hydrologic Modelling
- 4/24/92 - Review of EPAs Draft Comprehensive State Ground Water
- 5/28/91 - Final Report of EPAs Ground Water Task Force
- 4/13/90 - Re-establishment of the DOE Ground Water Strategy Work Group
- 3/5/90 - EPA Ground-Water Protection Principles and State-Federal Relation
- 4/4/88 - Department of Energy Groundwater Protection Paper
- 8/28/87 - Identification of Technical Guidance Related to Groundwater
- 2/04/87 - Groundwater Contamination at Department of Energy Facilities

Figure 3.3.9 Environmental Guidance Documents for the
Clean Water Act and Groundwater (2 of 2 pages)

Gopher Menu

- Search Clean Air Act data
- 12/30/94 - EPA Approval of Single Point Sampling using Shrouded Probe Tech.
- 12/28/94 - Final CAA Rule on Interim Requirements for Gas Detergent Additives
- 11/09/94 - CAA Hazardous Air Pollutant Sources, Early Reduction
- 10/31/94 - Final CAA Rule on Significant New Alternatives Policy Program
- 09/23/94 - Analysis of the Significant New Alternatives Policy Program
- 08/31/94 - EPA's List of Regulated Substances & Thresholds, Release Prevention
- 08/24/94 - EPA Stay of Requirements Related to Ozone-Depleting Substances
- 08/12/94 - Recommended Approaches to Management of Refrigerants
- 5/27/94 - INFORMATION - ANALYSIS OF THE EPA FINAL RULE RELATING TO STRATOSPHERIC OZONE PROTECTION, ACCELERATED PHASEOUT SCHEDULE
- 5/11/94 - INFORMATION — ANALYSIS OF THE CLEAN AIR FINAL RULE RELATING TO STRATOSPHERIC OZONE PROTECTION — LABELING REQUIREMENTS
- 3/30/94 - INFORMATION — FINAL CLEAN AIR ACT RULE ON SIGNIFICANT NEW ALTERNATIVES POLICY (SNAP) PROGRAM
- 3/25/94 - INFORMATION — ANALYSIS OF THE FINAL CLEAN AIR ACT REFRIGERANT RECYCLING RULE RELATING TO STRATOSPHERIC OZONE PROTECTION
- 03/22/94 - CY '93 Radionuclide Air Emissions Reports for DOE Sites
- 1/27/94 - INFORMATION — FINAL CLEAN AIR ACT RULE REQUIRING THAT FEDERAL
- 12/30/93 - ANNUAL REPORT ON PHASEOUT OF OZONE-DEPLETING SUBSTANCES
- 12/29/93 - INFORMATION — FEDERAL PROCUREMENT OF NON-OZONE-DEPLETING SUBSTANCES
- 11/3/93 - Draft Refrigerant Management Plan for Department of Energy
- 4/20/93 - ENVIRONMENTAL PROTECTION AGENCY RESPONSE TO DEPARTMENT OF ENERGY
- 4/6/93 - INFORMATION - SUMMARY OF STRATOSPHERIC OZONE PROTECTION
- 3/25/93 - Calendar Year 1992 Radionuclide Air Emissions Annual Reports
- 3/3/93 - INFORMATION - CLEAN AIR ACT FINAL LABELING REGULATIONS
- 1/4/93 - INFORMATION ON INITIAL LIST OF HAZARDOUS AIR POLLUTANT SOURCE
- 12/30/92 - Summary of Radionuclide Air Emissions
- 10/27/92 - SUBSTITUTES FOR OZONE-DEPLETING SUBSTANCES
- 10/19/92 - REPORT ON ALTERNATIVE CHEMICAL SUBSTITUTES AND TECHNOLOGIES
- 9/9/92 - CLEAN AIR ACT FINAL RULE - SERVICING OF MOTOR VEHICLE AIR
- 8/4/92 - 1991 Air Emissions Annual Report for Certain DOE Facilities
- 7/14/92 - CLEAN AIR ACT - ADDITIONAL GUIDANCE ON RECYCLING AND FUEL FLEET

Figure 3.5.9 Environmental Guidance Documents for the
Clean Air Act (1 of 3 pages)

- 7/2/92 - Submittal of DOE Annual Air Emissions Reports
- 6/25/92 - Certification of the Radionuclide NESHAPs Annual Air Emissions
- 6/18/92 - Information Copy - DOE Comments on Proposed Rule 40 CFR Part 82
- 5/22/92 - DISTRIBUTION OF DRAFT - AN INVENTORY OF HAZARDOUS AIR POLLUTANTS
- 5/7/92 - DISTRIBUTION OF PROPOSED LABELING REGULATIONS
- 5/1/92 - INFORMATION ON THE GENERAL PREAMBLE FOR THE IMPLEMENTATION
- 4/2/92 - CLEAN AIR ACT FINAL RULE - ADMINISTRATIVE ASSESSMENT OF
- 4/1/92 - CLEAN AIR ACT - SUPPLEMENTARY GUIDANCE ON RECYCLING OF
- 3/26/92 - CAP88-PC Available for DOE Use
- 3/12/92 - Field Offices Calendar Year 1991 Radionuclide Air Emissions
- 3/12/92 - CLEAN AIR ACT STATUTORILY-MANDATED COMPLIANCE DATES RELATED
- 1/29/92 - DISTRIBUTION OF NOTICE REQUESTING INFORMATION ON SUBSTITUTES
- 1/28/92 - CLEAN AIR ACT - EARLY EMISSION REDUCTIONS OF HAZARDOUS AIR
- 1/24/92 - New PC-based Version of CAP-88 Available for Review
- 1/14/92 - Air Quality Area Designations and Classifications
- 1/7/92 - Compliance with Emission Monitoring Requirements under 40 CFR
- 11/12/91 - Clean Air Act Amendments Workshop
- 11/5/91 - Compliance with Emission Monitoring Requirements under 40 CFR Par
- 10/16/91 - Compliance With Emission Monitoring Requirements under 40 CFR
- 10/10/91 - DISTRIBUTION OF PROPOSED REGULATIONS ON THE CLEAN FUEL FLEET
- 9/23/91 - REQUEST FOR INFORMATION CONCERNING HAZARDOUS AIR POLLUTANTS
- 9/19/91 - Clean Air Act (CAA) Reference Book
- 9/18/91 - CLEAN AIR ACT - ENHANCED MONITORING AND COMPLIANCE CERTIFICATION
- 9/18/91 - Asbestos NESHAP Training Requirements for On-Site Representative
- 9/11/91 - DISTRIBUTION OF PROPOSED REGULATIONS ON PROTECTION OF STRATOSPHER
- 8/14/91 - Information Copy — DOE Comments on the Preliminary Draft List
- 8/14/91 - Information Copy — Comments on the Environmental Protection
- 8/5/91 - Guidance on Achieving Compliance with the NESHAPs Emission
- 7/3/91 - DISTRIBUTION OF PRELIMINARY DRAFT LIST OF HAZARDOUS AIR
- 6/18/91 - DISTRIBUTION OF PROPOSED REGULATIONS FOR EARLY EMISSION
- 6/14/91 - DISTRIBUTION OF PROPOSED RULES CONTROLLING AIR RELEASES
- 5/20/91 - DISTRIBUTION OF PROPOSED RULE 40 CFR PART 70 - OPERATING PERMIT
- 5/20/91 - CLEAN AIR ACT - PROPOSED REGULATIONS FOR EARLY EMISSION
- 4/26/91 - Clean Air Act - New Source Review Program Transitional Guidance
- 4/26/91 - Calendar Year 1990 Annual Air Emissions Reports
- 4/15/91 - Clean Air Act Amendments of 1990
- 2/21/91 - 40 CFR Part 61 Standards for Radionuclide Emissions
- 7/12/90 - Los Alamos National Laboratory Air Emissions Compliance
- 6/15/90 - New Radionuclide National Emission Standards-40 CFR Part 61
- 1/26/90 - Promulgation of Radionuclide Emission Standards - 40 CFR Part 61

Figure 3.5.9 Environmental Guidance Documents for the
Clean Air Act (2 of 3 pages)

- 12/29/89 - Disparities in Regulations Governing Radionuclides
- 5/22/89 - DOE Comments on EPA Proposed Rulemaking, National Emission Standar
- 3/13/89 - State of Ohio Regulation of Radioactive Air Emissions
- 2/13/89 - EPAs Proposed Standards for Radionuclides under 40 CFR 61
- 11/23/88 - Draft Technical and Administrative Comments
- 11/10/88 - Submittal to the EPA on Potential Impact on DOE Operations
- 10/14/88 - Update on Status of EPA Clean Air Act Rulemakings Under 40 CFR Pa
- 10/5/88 - DOE Comments on EPAs Proposed Rulemaking for Benzene Emissions
- 9/22/88 - Draft DOE Comments Letter on EPAs Proposed Benzene Rule, 40 CFR
- 7/29/88 - Department of Energy Participation in the Environmental Protection
- 7/28/88 - EPA Proposal for a NESHAPS Standard for Benzene, July 20, 1988
- 5/23/88 - Developing Department of Energy (DOE) Recommendations
- 4/21/88 - Annual Radionuclide Air Emission Report - 40 CFR 61.94(c)
- 3/24/88 - Formation of DOE Steering Group to Coordinate Comments to EPA
- 1/29/88 - Environmental Protection Agency Critique of the Department of Ener
- 1/29/88 - Environmental Protection Agency Critique
- 8/31/87 - Advisory on the Authority of States to Regulate Radioactive Air
- 8/14/87 - State Of Washington Regulation of Radioactive Air Emissions
- 5/7/87 - Draft NESHAPS Approval Letter for Construction-Operation
- 4/24/87 - Annual Radionuclide Air Emission Report — 40 CFR 61.94(c)
- 4/6/87 - Compliance with 40 CFR 61, Subpart H
- 12/9/86 - Additional Guidance and Information for Implementing 40 CFR 61,
- 8/19/86 - Implementation of EPAs Radionuclide Air Emission Regulation,
- 5/8/86 - Guidance on 40 CFR 61 - Implementation for Radionuclide Emission
- 4/23/86 - Non-Applicability of 40 CFR 61.07, Application for Approval of
- 4/18/86 - Annual Radionuclide Air Emission Report - 40 CFR 61.94
- 4/9/86 - Request for Guidance on 40 CFR 61 - Implementation for
- 2/27/86 - Performance and Design Criteria for Monitoring Radionuclide Air
- 2/13/86 - Headquarters Guidance Regarding 40 CFR 61 Implementation.
- 12/13/85 - EPA Reporting Requirements for Radionuclides Regulated
- 8/5/85 - Radiation Standards for Protection of the Public

Figure 3.5.9 Environmental Guidance Documents for the
Clean Air Act (3 of 3 pages)

Gopher Menu

■ Search CERCLA data
■ 09/21/94 - Guidance on CERCLA Removal Actions
■ 08/22/94 - CERCLA Site Assessment Workbook
■ 07/19/94 - RCRA Corrective Action & CERCLA Remedial Action Reference Guide
■ 7/5/94 - Incorporating Ecological Risk Assessment into Remedial Investigation/Feasibility Study (RI/FS) Work Plans
■ 5/24/94 - Environmental Guidance on Reporting Releases of Hazardous Substances at DOE Facilities
■ 03/01/94 - The Off-Site Rule
■ 02/24/94 - A Guide to CERCLA Site Assessments
■ 02/01/94 - The Hazardous Ranking System
■ 12/21/93 - Environmental Guidance on CERCLA Remedial Investigation/Feasibility Study (RI/FS)
■ 11/9/93 - CERCLA Site Assessment Process Questions and Answers
■ 11/01/93 - Site Deletion from the NPL
■ 06/01/93 - Natural Resource Damages Under CERCLA
■ 06/01/93 - Site Inspections Under CERCLA
■ 05/05/93 - EO 12580: Superfund Implementation
■ 05/01/93 - Preliminary Assessments Under CERCLA
■ 3/29/93 - Nuclear Regulatory Commission (NRC) Low-Level Radioactive Waste
■ 2/10/93 - Superfund Reform
■ 12/16/92 - U.S. Environmental Protection Agency (EPA)
■ 12/7/92 - Request for Modification of DOE 5400.4 - Comprehensive
■ 10/26/92 - Site Assessments
■ 10/21/92 - EPA Superfund Accelerated Cleanup Model (SACM)
■ 9/30/92 - The Environmental Protection Agencys Revised Community Relations
■ 9/4/92 - Department of Energy CERCLA Orientation and RI-FS Workshop
■ 8/27/92 - Management of CERCLA Investigation-Derived Wastes
■ 8/5/92 - Natural Resource Trusteeship and Ecological Evaluation Workshop
■ 7/23/92 - DOE-EPA Environmental Release Reporting Workshop
■ 6/5/92 - CERCLA INFORMATION BRIEF - EH-231-012/0692, JUNE 1992
■ 5/22/92 - Proposed Procedures for Contract Laboratory Program
■ 5/18/92 - Superfund
■ 5/18/92 - Hazardous Waste Remediation
■ 5/15/92 - Notice of Proposed Rulemaking on Reportable Quantity
■ 5/15/92 - Natural Resource Trusteeship and Ecological Evaluation
■ 5/14/92 - EPA CERCLA Orientation Workshop
■ 4/22/92 - 1992 Compendium of Superfund Program Publications and Report
■ 4/3/92 - Department of Energy (DOE) Natural Resource Damage
■ 3/24/92 - Applicable or Relevant and Appropriate Requirements
■ 3/20/92 - EPAs Regional Site Assessment Training Workshops and EPA
■ 3/13/92 - Superfund Annual Report to congress
■ 3/11/92 - Preliminary Assessment-Site Information CERCLIS Update
■ 2/28/92 - Judgement Affecting Department of Energy Facilities
■ 1/10/92 - EPA Preliminary Assessment-Site Investigation (PA-SI) Workshop
■ 1/7/92 - Guidance for Performing Preliminary Assessments Under CERCLA
■ 1/5/92 - CERCLA INFORMATION BRIEF - EH-231-011/0192, JANUARY 1992
■ 12/17/91 - EPA Draft - Guidance on Use of Institutional Controls at Superfu

Figure 3.6.12A Environmental Guidance Documents
for CERCLA (1 of 3 pages)

- 11/14/91 - Comprehensive Environmental Response, Compensation and Liability
- 11/7/91 - Department of Energy 1991 Comprehensive Environmental Response,
- 11/4/91 - Integrity of Environmental Analytical Data
- 11/1/91 - DOE Remedial Investigation-Feasibility Study (RI-FS) Workshop
- 10/18/91 - CERCLA Orientation Workshop
- 9/16/91 - Judgement Affecting Certain CERCLA Docket Facilities
- 8/8/91 - A Comparative Analysis of Remedies
- 8/8/91 - 1990 Comprehensive Environmental Response, Compensation
- 8/6/91 - Improving Remedy Selection
- 7/17/91 - Report of the EPA Superfund Management Review
- 7/12/91 - Consolidated DOE Response to Notice of Proposed Rulemaking On CER
- 6/17/91 - CERCLA Federal Agency Hazardous Waste Compliance Docket
- 6/4/91 - Guidance for Natural Resource Trusteeship and Ecological
- 5/22/91 - Guidance on Reporting Requirements for Continuous Releases of
- 5/15/91 - Superfund Program Analysis—Lessons Learned
- 5/15/91 - Proposed Revisions to CERCLA Natural Resource Damage Assessments
- 2/19/91 - CERCLA activities and DOE D&D Activities
- 1/16/91 - DOE Comments on Draft Proposed Rule for CERCLA Natural Resource
- 11/9/90 - Orientation Course on The Revised Hazard Ranking System (HRS)
- 9/18/90 - The Superfund Remedial Action Decision Process
- 9/14/90 - Superfund Program
- 9/7/90 - Superfund from the Industry Perspective
- 9/7/90 - Making Superfund Work
- 8/28/90 - Understanding Superfund
- 8/28/90 - Missed Statutory Deadlines Slow Progress
- 8/17/90 - Fact Sheets covering two Reports on Environmental Protection
- 7/30/90 - Fact Sheet, 10 Superfund Case Studies
- 7/27/90 - Federal Real Property Transfer Rule
- 7/05/90 - Catalog of Applicable Relevant and Appropriate Requirements
- 5/29/90 - Risk Assessment Guidance for Superfund - Training
- 5/29/90 - Interim Guidance on Reportable Quantities and Radionuclide
- 4/23/90 - Revised National Contingency Plan
- 3/28/90 - Applicability of Land Disposal Restrictions to RCRA and CERCLA
- 2/23/90 - Three Upcoming Workshops Covering Reporting Requirements
- 2/14/90 - DOE Comments on EPA Proposed Policy and Guidance Document
- 2/12/90 - Transmittal of Report - Comparative Review of U.S. DOE
- 1/22/90 - Applicability of Proposed Conflict of Interest Policy at Pantex
- 1/10/90 - Final Reportable Quantities of Radionuclides and Rules for
- 1/4/90 - Remedial Investigation-Feasibility Study (RI-FS) Workshop
- 12/15/89 - Fact Sheets
- 12/14/89 - Draft Guidance on Reportable Quantity (RQ) Notification
- 11/8/89 - Issuance of DOE 5400.4
- 5/08/89 - DOE Comments on the Proposed Revision to the National Contingenc
- 2/22/89 - Issuance of DOE 5400.3
- 2/07/89 - Federal Facility Agreement under CERCLA for the Monticello Vicin
- 2/06/89 - Consent Decrees for the Feed Materials Production
- 2/01/89 - Hazard Ranking System Briefing
- 1/30/89 - RI-FS Guidance
- 1/23/89 - Request for Comments on the Proposed Revisions to the Hazard
- 1/16/89 - Summary of CERCLA Preliminary Assessment Submittals
- 12/12/88 - Federal Facility Agreement under the Comprehensive Environmental
- 10/3/88 - Disposition of Issues on Draft DOE 5400.YY, COMPREHENSIVE

Figure 3.6.12A Environmental Guidance Documents
for CERCLA (2 of 3 pages)

- 9/23/88 - Payments to States for CERCLA Response Costs
- 8/24/88 - Request for Comments on the Proposed Rule for Reporting Exemption
- 8/16/88 - DOE Notice 5400.4 - Integration of Environmental Compliance
- 8/15/88 - Federal Agency Hazardous Waste Compliance Docket Update
- 8/5/88 - Contractor Conflict-of-Interest Issues Under the Comprehensive
- 6/24/88 - Remedial Investigation-Feasibility Study (RI-FS) Work Plan
- 5/31/88 - Agreement with the Environmental Protection Agency — Model
- 4/7/88 - DOE-RL Request for Guidance on Funding of Regulatory Oversight
- 3/7/88 - Revision of DOE 5480.14, Comprehensive Environmental Response,
- 1/22/88 - Follow-up to PUREX Decladding Waste Issues
- 10/22/87 - Revisions to the National Contingency Plan (NCP)
- 10/2/87 - Submission of Information for National Priority List Scoring
- 9/9/87 - Interim Guidance on Applicable or Relevant and Appropriate
- 9/8/87 - Interim Guidance on Administrative Records
- 3/6/87 - Executive Order for Superfund Implementation
- 2/26/87 - DOE Participation on the Regional Response Teams
- 1/15/87 - Planned Revision of DOE Order 5480.14, Comprehensive Environmenta
- 10/23/86 - Review of the Comprehensive Environmental Response, Compensation
- 6/13/86 - Strategy for Requesting EPA Not to List the Lawrence Livermore
- 5/14/86 - Superfund Money for Department of Energy Cleanups

---

Figure 3.6.12A Environmental Guidance Documents
for CERCLA (3 of 3 pages)

Gopher Menu

■ Search Resource Conservation and Recovery Act data
■ 09/30/94 - Selecting Compliant Off-Site Hazardous Waste TSD Facilities
■ 5/11/94 - Hazardous Waste Identification Rule (HWIR)
■ 5/4/94 - Cancellation of DOE 5400.3
■ 05/01/94 - RCRA CORRECTIVE ACTION HUMAN HEALTH ASSESSMENT
■ 04/04/94 - Transportation of RCRA Hazardous Wastes
■ 03/01/94 - CORRECTIVE ACTION MANAGEMENT UNITS AND TEMPORARY UNITS
■ 03/01/94 - RCRA CORRECTIVE ACTION DEFINITIONS UNDER SUBPART F AND PROPOSED SUBPART S
■ 01/01/94 - HAZARDOUS SUBSTANCE USTs RCRA SUBTITLE I UNDERGROUND STORAGE TANKS
■ 01/01/94 - PETROLEUM USTs RCRA SUBTITLE I UNDERGROUND STORAGE TANKS
■ 11/26/93 - Transfer of Nuclear Weapon Components to the Private Sector for
■ 11/23/93 - RCRA Administrative Hearing and Appeals Process
■ 11/01/93 - Ground-Water Monitoring Under RCRA
■ 08/08/93 - Inspections of RCRA Container Storage Areas
■ 07/01/93 - Requirements for Satellite Accumulation Areas
■ 07/01/93 - RCRA HW Container Labeling, Marking, and Placarding Requirements
■ 06/01/93 - The "Derived-from" Rule under RCRA
■ 05/05/93 - Identifying and Classifying an UST
■ 05/01/93 - Deferred USTs: RCRA Subtitle I, USTs
■ 05/01/93 - Excluded USTs, RCRA Subtitle I, USTs
■ 2/16/93 - Environmental Guidance on Preparing a Waste Analysis Plan for RC
■ 01/01/93 - RCRA Air Emission Standards for HW TSDF Process Vents
■ 01/01/93 - RCRA Air Emission Standards for HW TSDF Equipment Leaks
■ 11/18/92 - Hazardous Waste Identification Rule (HWIR) Notice of Proposed
■ 11/12/92 - Impact of the Federal Facilities Compliance Act on Outstanding R
■ 10/30/92 - Interim Final Hazardous Soil Case-By-Case Capacity Variance
■ 10/22/92 - Identification and Listing of Hazardous Waste Used Oil - Final
■ 9/24/92 - Definitions of Solid and Hazardous Waste Workshops
■ 9/1/92 - RCRA INFORMATION BRIEF - EH-231-0181/0992, SEPTEMBER 1992
■ 8/14/92 - Compendium of Documents Relevant to RCRA Corrective Action
■ 8/3/92 - Definitions of Solid and Hazardous Waste Under the Resource
■ 8/3/92 - Consolidated Departmental Response to the Hazardous Waste
■ 6/19/92 - Environmental Guidance on Regulated Underground Storage Tanks
■ 6/19/92 - Consolidated DOE Response to NRC-EPA Draft Guidance
■ 5/20/92 - Resource Conservation and Recovery Act (RCRA) Reference Book
■ 5/20/92 - Resource Conservation and Recovery Act (RCRA) - Mixture
■ 5/15/92 - RCRA Generator Biennial Report
■ 5/8/92 - Regulatory Bulletin - Solid Waste Disposal Facility
■ 5/6/92 - Liners and Leak Detection Systems for Hazardous Waste Land
■ 5/1/92 - RCRA INFORMATION BRIEF - EH-231-011/0592, MAY 1992
■ 4/9/92 - Consolidated Department of Energy (DOE) Response
■ 3/26/92 - DOE-EPA RCRA Corrective Action Workshop for Federal
■ 3/3/92 - The Resource Conservation and Recovery Act (RCRA) Mixture

Figure 3.6.12B Environmental Guidance
Documents for RCRA (1 of 4 pages)

- ■ 2/12/92 - Timing of Surface Impoundment Retrofitting Under the Land Disposa
- ■ 2/3/92 - Amendments to Interim Status Standards for Downgradient Groundwat
- ■ 1/27/92 - Nonhazardous Waste: Environmental Safeguards for Industrial Facil
- ■ 1/9/92 - Proposed Rule on Land Disposal Restrictions (LDR) For Newly Liste
- ■ 12/20/91 - EPA RCRA Orientation Workshop
- ■ 12/9/91 - Consolidated Department of Energy (DOE) Response to VOC Organic
- ■ 11/27/91 - Limited Progrss in Closing and Cleaning Up Contaminated Faciliti
- ■ 11/14/91 - Land Disposal Restrictions — Potential Treatment Standards
- ■ 11/14/91 - Consolidated Department of Energy (DOE) Response to Supplemental
- ■ 10/29/91 - Land Disposal Restrictions (LDR) Notice — Potential Treatment
- ■ 10/3/91 - Used Oil Supplemental Notice of Proposed Rulemaking
- ■ 9/30/91 - Underground Storage Tank Technical Requirements, Final Rule
- ■ 9/13/91 - Extension of Comment Period for the Hazardous Waste Treatment
- ■ 9/3/91 - DOE-EPA RCRA Corrective Action Workshop for Federal Facilities
- ■ 8/7/91 - Collection of Hazardous and Mixed Waste Data on New Toxicity
- ■ 8/5/91 - Storage of Land Disposal Restricted Mixed Waste
- ■ 8/2/91 - Proposed Standards for Volatile Organic Emissions from Hazardous
- ■ 7/30/91 - RCRA Definition of Solid and Hazardous Wastes Training Course
- ■ 7/30/91 - Burning of Hazardous Waste in Industrial Furnaces (BIF) - Final
- ■ 7/17/91 - The RCRA Implementation Study
- ■ 7/16/91 - Videotape of Training Session on RCRA Toxicity Characteristic
- ■ 7/8/91 - Mixed Waste and Materials Management Issues
- ■ 6/4/91 - Land Disposal Restrictions Advance Notice of Proposed Rulemaking
- ■ 5/01/91 - Synopsis of Final Rule - Burning of Hazardous Waste in Industrial
- ■ 4/8/91 - Proposed Amendments to Interim Status Standards for Downgradient
- ■ 4/3/91 - Contaminated Soil and Debris Wastes Generated from DOE Environmen
- ■ 3/27/91 - Toxicity Characteristic Training for Waste Management and Regulat
- ■ 3/21/91 - Hazardous Wast: Status and Resources of EPAs Corrective Action P
- ■ 3/14/91 - Applicability of the Toxicity Characteristic to Mixed Waste
- ■ 2/27/91 - Suspension of Resource Conservation and Recovery Act (RCRA) Toxic
- ■ 2/12/91 - Amendments to Interim Status Standards for Down-gradient Groundwa
- ■ 1/3/91 - RCRA Subpart S Corrective Action Proposed Rule
- ■ 12/21/90 - Toxicity Characteristics Training for Waste Management and Regul
- ■ 11/23/90 - Harzardous Waste: EPAs Generation and Management Data Need Furth
- ■ 10/17/90 - Environmental Protection Agency Notice on Waste Minimization
- ■ 9/18/90 - The Environmental Protection Agencys (EPA) GUIDEBOOK FOR
- ■ 7/20/90 - Review of RCRA Subpart S Corrective Action Proposed Rule.
- ■ 7/20/90 - DOE Corrective Action Workgroup Meeting on Subpart S Corrective
- ■ 7/9/90 - Final Rule Regarding Land Disposal Restrictions
- ■ 7/3/90 - Westinghouse Corporate Concerns Regarding SEN-22 (RCRA Permit
- ■ 7/2/90 - Land Disposal Restriction Videoconference
- ■ 6/27/90 - WASTE REDUCTION POLICY STATEMENT
- ■ 6/18/90 - Waste Testing and Quality Assurance Symposium, July 16-20, 1990
- ■ 6/05/90 - Negotiations Skills Development Training Course
- ■ 5/25/90 - Alabama Ban on Out-of-State Hazardous Waste
- ■ 5/24/90 - Delay of Closure Period for Hazardous Waste Management Facilitie
- ■ 5/21/90 - Land Disposal Restrictions for Third Scheduled Wastes
- ■ 5/18/90 - List of Outstanding Notices of Violation and Warning Letters
- ■ 5/11/90 - Emission Controls for Hazardous Waste Incinerators
- ■ 5/8/90 - Underground Storage Tanks, Technical Requirements
- ■ 5/04/90 - Clarification of the Definition of Designated Facility under RCR

Figure 3.6.12B Environmental Guidance
Documents for RCRA (2 of 4 pages)

- 4/24/90 - RCRA Mixed Waste Meeting
- 4/23/90 - Comments to OMB on RCRA Subpart S Corrective Action Draft
- 4/20/90 - The Environmental Protection Agencys Proposed Conditional
- 4/13/90 - Teleconference on Third Rule
- 4/9/90 - Toxicity Characteristic Final Rule
- 3/28/90 - Applicability of Land Disposal Restrictions to RCRA and CERCLA Gr
- 3/22/90 - Distribution of Hazardous Material-Waste Contingency Plans
- 3/21/90 - Savannah River Site (SRS) Compliance Issues
- 3/20/90 - Data on Calcination and Storage as BDAT for ICPP HLLW
- 3/08/90 - Land Disposal Restrictions Training via Satellite
- 2/17/90 - Amendments to EPA Manual on Test Methods for Evaluating Solid Wa
- 2/15/90 - DOE RCRA Compliance Issues Workgroup
- 2/13/90 - Proposed Land Disposal Restrictions for Third Scheduled
- 2/6/90 - 1990 Inventory of Federal Hazardous Waste Activities
- 2/1/90 - Environmental Protection Agency (EPA) Land Disposal Restrictions
- 1/30/90 - Management of Solvent Laden Rags as RCRA Hazardous Waste
- 1/8/90 - Inventory of Federal Hazardous Waste Activities under Section 301
- 1/2/90 - Corrective Action Plan, Interim Measures Guidance, National
- 12/21/89 - Workgroup on Environmental Protection Agency Proposed Subpart S R
- 12/20/89 - The Environmental Protection Agencys (EPA) 1989 Waste Minimizatio
- 12/15/89 - Fact Sheets
- 12/6/89 - Issuance of DOE 5400.3
- 11/29/89 - RCRA Land Disposal Restrictions (LDR) Summary Booklet
- 11/21/89 - Implementation of the Medical Waste Tracking Act
- 7/12/89 - The Environmental Protection Agencys Draft Guidance to Hazardous
- 6/21/89 - RCRA Land Disposal Restrictions
- 6/01/89 - LDR Strategy
- 4/21/89 - Extension of Manifest Espiration Date
- 4/11/89 - Proposed Land Disposal Restrictions for Second Third Scheduled
- 4/03/89 - EPA Final Rulemaking on Permit Changes to Interim Status
- 3/20/89 - The Environmental Protection Agencys Proposed Pollution
- 3/13/89 - Environmental Protection Agency Proposal to Broaden Restrictions
- 3/8/89 - Compendium of Office of Research and Development (ORD) and Office
- 1/24/89 - Guidance for Complying with the Notification-Certification
- 1/18/89 - Proposed Land Disposal Restrictions for Second Third
- 11/22/88 - Modifications of Hazardous Waste Management Permits
- 11/14/88 - Review of RCRA Land Disposal Restrictions First-Third Final Rule
- 10/28/88 - Final EPA Underground Storage Tank Regulations, Subtitle I, Soli
- 10/13/88 - Review of Draft Land Disposal Restrictions
- 10/05/88 - Land Disposal Restrictions (LDR) Facilities Assessment
- 8/29/88 - Environmental Protection Agencys Significant Non-compliers List
- 8/19/88 - RCRA Hammer Dates for Permit Applications
- 6/15/88 - Environmental Protection Agency Memoranda On RCRA Issues
- 6/2/88 - Proposed Changes to the RCRA Toxicity Characteristic
- 5/16/88 - EPA Elevation Process for RCRA Federal Facility Compliance Agreem
- 4/26/88 - Land Disposal Restrictions Summary Booklet
- 4/20/88 - Signatures on the RCRA Part A Permit
- 2/29/88 - Federal Agencey Hazardous Wate Compliance Docket
- 12/24/87 - Signatures on RCRA Permit Applications
- 11/24/87 - Regulation of Single Shell Tanks at Hanford
- 10/22/87 - Inventory of Federal Hazardous Waste Sites

Figure 3.6.12B Environmental Guidance
Documents for RCRA (3 of 4 pages)

■ 9/21/87 - Statistical Methods for Evaluating Groundwater Monitoring data
■ 8/10/87 - Submittal of RCRA Part A Permit Applications for Hanford
■ 8/6/87 - Resource conservation and Land Ban Information
■ 8/6/87 - Comments on underground storage tanks
■ 7/28/87 - Consent order and Compliance agreement
■ 7/8/87 - Contingency Plan for Compliance with California List Land Dispos
■ 7/07/87 - U.S. Environmental Protection Agency and State of Tennessee Reso
■ 6/11/87 - Submittal of RCRA Part A Permit Applications for Mixed Waste Trea
■ 6/11/87 - Federal Agency Hazardous Waste Compliance Docket
■ 6/10/87 - Land Disposal Restrictions—The California List
■ 6/2/87 - Petitions to Delist Hazardous Waste
■ 4/22/87 - Technical Resource Document for Obtaining variances from the Seco
■ 2/19/87 - EPAs List of Significant Non-Compliers for Fiscal Year 1987
■ 2/4/87 - Revision of the RCRA Subtitle
■ 11/7/86 - Byproduct policy review
■ 8/15/86 - Variance Guidance Document
■ 8/12/86 - Information request from EPA
■ 7/23/86 - Request for information from EPA
■ 6/24/86 - Policy on Off-Site Treatment, Storage, and Disposal of Nonradioa
■ 5/23/86 - DOE-Wide Policy on Off-Site Treatment, Storage, and Disposal of N
■ 4/18/86 - EPA Groundwater Monitoring Technical Guidance
■ 4/11/86 - Reporting Presence of Underground Storage Tanks Containing Radio
■ 3/14/86 - Draft Environmental Protection Agency Memorandum on Federal Facil
■ 3/7/86 - Reporting Presence of Underground Storage Tanks on Department of
■ 2/28/86 - Preparation of Annual Site Environmental Reports for Calender Yea
■ 1/2/86 - Inventory of Department of Energy Hazardous Waste Facilities-Sect
■ 11/27/85 - Inventory of Department of Energy Hazardous Waste Facilities
■ 10/28/85 - RCRA Gound-Water Monitoring Certification
■ 9/16/85 - Supplemental Interim Guidance for Compliance with the Resource Co
■ 7/2/85 - Guidance for Compliance with the New Waste Minimization Requireme

Figure 3.6.12B Environmental Guidance
Documents for RCRA (4 of 4 pages)

Gopher Menu

■ Search Toxic Substance Control Act data
■ 11/1/93 - Environmental Guidance on the Management of Polychlorinated
■ 11/24/92 - Remedial Actions at DOE Environmental Restoration Sites
■ 7/21/92 - Toxic Substances Control Act (TSCA) Reference Book
■ 9/4/91 - Consolidated DOE Response to Advance Notice of Proposed Rulemakin
■ 8/27/91 - Consolidated DOE response to Advanced Notice of Proposed Rulemaki
■ 6/14/91 - Advance Notice of Proposed Rulemaking (ANPRM) on PCB Disposal
■ 5/23/91 - Advanced Notice of Proposed Rulemaking, Comprehensive Review
■ 2/5/90 - PCB Notification and Manifesting Final Rule
■ 10/28/88 - Proposed Rule on Polychlorinated Biphenyls Notification
■ 11/20/87 - Toxic Substances Control Act Complaint Against Rockwell
■ 8/04/87 - Proposed Agreements Between the Environmental Protection
■ 6/1/87 - PCB SPILL SAMPLING MANUALS
■ 4/17/87 - Environmental Protection Agency Policy Statement
■ 9/05/86 - Negotiations with the Environmental Protection Agency
■ 8/15/86 - TSCA-PCB Violations

Figure 3.8.7 Environmental Guidance Documents for TSCA (1 of 1 page)

Index

N

O

P

R

S